MRE

Materials Research and Engineering
Edited by B. Ilschner
Volume 1

D. G. Altenpohl

Materials in World Perspective

Assessment of Resources, Technologies and Trends for Key Materials Industrie

In Collaboration with T. S. Daugherty

With Contributions by M. B. Bever, J. P. Clark, H. Eldag, G. Friese, P. Kelterborn, I. Reznik, D. Spreng, F. R. Tuler

With 33 Figures

Springer-Verlag Berlin Heidelberg NewYork 1980

Dr. rer. nat. DIETER GUSTAV ALTENPOHL
Vice President Technology, Swiss Aluminium Ltd., Zurich
Visiting Professor of Materials Science at the University of Virginia

Dr. rer. nat. BERNHARD ILSCHNER
o. Professor, Institut für Werkstoffwissenschaften der Universität Erlangen-Nürnberg

Library of Congress Cataloging in Publication Data.
Altenpohl, Dietrich. Materials in world perspective.
(Materials research and engineering ; v. 1)
Bibliography: p. Includes index.
1. Materials. II. Title. III. Seires: Materials research
and engineering ; v. 1.
TA 403. A 517 338.4'767 80-13301

ISBN-13:978-3-642-81455-6 e-ISBN-13:978-3-642-81453-2
DOI: 10.1007/978-3-642-81453-2

2362/3020 – 543210

Editor's Preface

This is the first book in a new series – "Materials Research and Engineering" – devoted to the science and technology of materials. "Materials Research and Engineering" evolves from a previous series on "Reine und Angewandte Metallkunde" ("Pure and Applied Metallurgy"), which was edited by Werner Köster until his eightieth birthday in 1976.

Although the present series is an outgrowth of the earlier one, it should not and cannot be regarded as a continuation. There had to be a shift of scope – and a change in presentation as well. Metallurgy is no longer an isolated art and science. Rather, it is linked by its scientific basis and technological implications to non-metallic and composite materials, as well as to processes for production, refining, shaping, surface treatment, and application. Thus, the new series, "Materials Research and Engineering", will present up-to-date information on scientific and technological progress, as well as on issues of general relevance within the engineering field and industrial society.

Premiering the new series, the present book by Dieter Altenpohl gives the reader a very general outlook, in fact, a position analysis of materials and the materials industry within the framework of our contemporary technological environment. It ventures, moreover, to forecast the changes affecting this pattern in a dynamic, interdependent world. This may be an unusual way to start a scientific series – it is believed, nevertheless, to be an appropriate one.

In recent years, new aspects have been introduced to the scientific community – sometimes embarrassing, sometimes stimulating, in any case important: "limits-to-growth", "resources", "responsibility" and others. While creative thinking is in demand more than ever before, a new consciousness of the consequences of boundless creativity has emerged. Altenpohl's book intends to focus attention on this new responsibility, on the interrelations and interactions between different fields of technology and between different areas of our world. For this reason, its topic is felt to be appropriate as setting a stage – which will be followed by titles that represent the broad field of technological innovation and scientific progress.

The importance of the topic is matched by the competence of the author, having at his disposition a broad experience and world-wide reputation both as a metallurgical scientist and a critical analyst of scenarios for a future world. In particular, he is the author of Vol. 19 of the precursor series, a comprehensive monograph on "Aluminium and Aluminium Alloys", as well as other books. It is anticipated, therefore, that the present book will gain the attention of a wide audience, and will make, together with the subsequent scientific and technological publications within the series, a worthwhile contribution to the development of better engineering materials for responsible use in a human society.

Erlangen, Germany, January 1980 *Bernhard Ilschner*

Foreword

During the 1960's when I was chairman of the US House of Representatives Subcommittee on Science Research and Development, the basic characteristics of American science policy were a natural subject for examination. Our country, accustomed for so long to a surplus of resources and the unlimited uses thereof, began to recognize not only that our resources were in short supply, but that our habits about them, so deeply ingrained in our economic structure, would not be easily changed. One of the tools we originated for the examination of that issue, as well as numerous others affecting the environment and the relationship between producers and consumers, all of which are so intrinsically involved, was the concept of technology assessment, which now reflects itself in so many formal governmental and non-governmental structures.

For these past 15 or 20 years as a result, our country has been seriously engaged in a process of adjustment from what many have classified as a "throw-away economy" where a constantly increasing standard of living for our people has depended on a productive cycle which has tended towards the excessive use and depletion of our material resources. As we have adjusted our thinking to a new set of requirements, we have set in motion a series of conflicts which are not well understood or easily resolved. Simply stated – as a technologically oriented society, we have traditionally exploited new technologies as soon as they were available and of late we have wanted to get rid of them as soon as they have been perceived to be harmful. Since these applications have become such an important part of our daily lives, it is difficult to make such changes without stirring up conflicts which often have traumatic social and economic effects.

During the last decade, such fundamental changes have been particularly related to the materials industries world-wide. These changes can be found in all segments of the materials cycle from the basic feedstocks, like minerals and ores, to the energy-intensive, metallurgical and processing industries on to the final product, such as the automobile and other material-intensive consumer goods.

There are many different elements of these changes. Significant among a host of activities was the study of the Club of Rome on "The Limits to Growth" and the burgeoning environmental movement of the late 1960's which still continues unabated to this day. From both resulted a public awareness that materials rank with energy high on the list of the basic resources of mankind. As a backbone of our civilization, materials should therefore be utilized and safeguarded in connection with a careful assessment of all potential impacts during the total life cycle of a material.

The materials industry in highly developed countries is entering a new era. Less materials and less energy per unit of gross national product will be used and this trend has already clearly developed in the last few years. In less developed countries the demand for appropriate technology is clearly recognized as well as the demand of processing their indigenous raw materials within the country to add value as much as possible.

There are many other trends to be observed, such as the need for the automotive industry to reduce weight of its vehicles to save energy. This will create an intensive process of substitution of materials and will favor recyclable materials.

Only a total system approach can provide decision-making tools in this complex process of change. The assessment of eligible technologies, materials and systems is now of utmost importance, and high attention has been paid to this, since 1972, by the Office of Technology Assessment and other institutions. In the recent few years, a methodology for assessment of this type has been developed and it has been amply demonstrated in numerous case histories as a useful instrument. This book now applies the methodology of technology assessment to the main materials industries in their dramatic set of a changing environment and new uncertainties. It applies this methodology in a broad and general way that is comprehensible and of value to the general reader as well as to the specialist.

The necessity for an overall management of our materials resources towards more efficient use and recycling is unquestioned. This book presents an introduction into the changes which are ongoing, then describes the six key materials industries, and afterwards explains how technology assessment can be carried out. This book is particularly timely and will serve as background material for decision-makers not only in governmental organizations but also those in industry, finance and academia. It will be of interest also to that segment of the public which is interested in the interaction between materials, energy and environment.

E. Q. Daddario, Washington, D.C.

Past Chairman and President
of AAAS

Former Director of The Office
of Technology Assessment

Author's Preface

Our book covers the present status and ongoing developments in the materials industry in highly developed countries as well as in resource-rich developing countries. It deals with the entire material's cycle. For the first time, this book interrelates resources, technologies, innovations and trends in key industries producing or using materials.

There are important changes taking place from the mine through the refining stage, on to finished products which are materials intensive. These, and techno-economic trends already recognizable today, are outlined. Of specific importance are the impacts which a given material or technology has on the total environment, including energy, ecology and social acceptability.

These impacts can be evaluated by careful technology assessment which should be integrated into the technology planning by the materials industry for its further healthy development.

We take the approach of an integrator to reveal correlations between the materials industry and key issues in its total environment. In doing so, the text was intentionally kept easily understandable and serves as an introduction for a wide range of readers with quite different backgrounds.

In writing this book, I received great help from many of my colleagues in Europe and abroad. My special thanks go to the Honorable *Emilio Daddario,* who created my interest in the subject of technology assessment within the materials industry; to Professor *B. Ilschner,* who made valuable suggestions regarding the content of the book; further to Dr. *Frank Huddle,* Senior Materials Specialist of the Congressional Research Service and to Dr. *S. Victor Radcliffe,* Senior Fellow, Resources for the Future, both in Washington D.C., for stimulating discussions. I wish to express my deep appreciation to the eight authors of specific topics and further to my associate *T.S. Daugherty* for his excellent cooperation and interest in the book.

Last, but not least, I wish to thank the University of Virginia, School of Engineering and Applied Science, and its Department for Materials Science for the opportunity to exchange views with members of their faculty and further to the general management of Swiss Aluminium Ltd. for their support of this book.

D. G. Altenpohl, Zurich

Contents

I Role of Materials in the World Economy

1 Materials and Man's Needs

Civilization has passed through several "ages" each identified with one material. These ages include the Stone Age, the Bronze Age and the Iron Age. Now mankind is in the "Multi-materials Age". A good illustration of the role materials play in day-to-day living is the ordinary domestic telephone. It contains 42 of the 92 naturally occurring elements in its components.

The widespread use of materials in industrial activity has resulted in a concern over their availability and given rise to studies about materials policy issues. Several reports have been published in the USA and Europe. Two of the more recent and important reports are the COSMAT Report (by the Committee on the Survey of Materials Science and Engineering) and the COMRATE Report (by the Committee on Mineral Resources and the Environment), published by the National Academy of Sciences (NAS) in the USA.

The 1974 COSMAT Report "Materials and Man's Needs", obtained its basic information from approximately 1000 questionnaires which were sent out in North America to many different institutions dealing with materials, from basic resources to finished products. Two important concepts were set forth in the report.

- Materials should be dealt with systematically as part of an overall materials cycle – from their extraction from the earth; through processing, design, and manufacture; use; and reuse or disposal back into the earth.
- Materials are only part of the basic triad of materials-energy-environment which should be treated as a whole in dealing with man's relationship to natural resources and effective national materials policy.

The COSMAT Report made 24 recommendations that included:

- a call for a definitive materials policy as it related to energy and the environment;
- support of materials science and engineering;
- strengthened mangement of materials information;
- measures to exploit renewable materials and recycling;
- more attention to the international aspects of materials.

The report "Mineral Resources and the Environment", published by the COMRATE in 1975, dealt with four interrelated materials problem areas:

– materials conservation through technology;
– estimation of mineral reserves and resources;
– the implication of minerals production for health and the environment;
– the demand for fuel and mineral resources.

An important theme of this report was that no aspect of materials policy could be considered in isolation.

Like the earlier COSMAT Report, COMRATE emphasized the importance of the materials cycle, the materials-energy-environment triad and the need for a systems approach to materials policy. COMRATE also had much to say about the incompleteness and unreliability of materials information. Many of its recommendations were directed at improvements in materials information systems.

These studies have pointed out that, today, there is a need for overall, creative, and wise management of the world's materials resources toward a closed cycle of efficient use and reuse.

This does not apply solely to industrialized countries. The developing countries or some of the emerging nations, rich in natural resources, have corresponding needs and problems also.

For these countries the problem of how to develop their natural resources: specifically ores, fossil fuels or biomass, by added value or just for export gives rise to a number of questions which need to be resolved in the coming decades.

Last but not least, we must not overlook the material needs of those countries, sometimes grouped as the Fourth World. This group consists of approximately 30 countries and they are located mainly in Africa, Southeast Asia and the Indian subcontinent. These countries are called "least developed" by the United Nations Organization. They are not rich in resources but still have an urgent need for materials.

2 Definition of Involved Materials

There are three main groups of raw materials to be considered for the materials industry.

1) Minerals and ores.
2) Fossil fuels (coal, oil and natural gas).
3) Biomass (e.g., wood, cotton, sugar cane, etc.).

But, first let us briefly look at some definitions of materials. Economists view materials as all inputs to industry. These include "energy materials" (oil, coal, gas, and nuclear fuel) as well as "industrial materials" (metals, cement and other construction materials, glass, lumber, paper, textiles, chemicals including plastics).

The Encyclopedia of Materials Science and Engineering* defines materials already somewhat narrower as physical matter which is manipulated and used by man, generally without a major chemical conversion.

The Congressional Research Service of the US Library of Congress, as an attempt to define materials functionally for legislative purposes, uses the following definition:

> "Materials means natural resources intended to be utilized by industry in the production of goods, with the exclusion of food and of energy fuels used as such."

In these definitions, the term "material" is used for minerals or ores as well as for primary metals, plastics, wood and cement. We will try to differentiate between these two groups by calling the first group "raw materials" which are inputs to the materials industry and, the second group "basic materials" which are outputs from the materials industry. Basic materials are predominantly of primary origin, but there is an increasing supply contribution from secondary sources due to recycling.

Since the materials industry is the main concern of this book we will describe those materials from which durable goods and packaging are produced. Therefore, food and drugs, textiles, and chemicals other than plastics, are excluded from this book. Thus, we selected the rather coherent group of engineering materials which are important in today's civilization to form infrastructures, provide housing, manufacture vehicles, machines, electrical and electronic equipment, or to make packages. To provide a complete picture, we deal with the extraction, processing and use of materials which are of utmost importance for the provision of industrial goods.

3 Setting the Stage

Raw materials are not uniformly distributed around the world. Economically useful concentrations of specific minerals can range from abundant in some countries to nonexistent in others.

Today, the highest consumption of materials is in countries of the North Temperate Zone. However, there is no relation between a country's rate of materials consumption and its raw material resources or reserves. A high level of industrialization tends to go hand-in-hand with dependence on imports of raw and and basic materials.

* In preparation, Pergamon Press.

Because minerals, energy and organic materials are essential to the manufacturing sector of industry it is necessary to have a balance in their availability.

3.1 The Total Materials Cycle

The starting point for our considerations is the total materials cycle. Figure 1 shows the flow of materials from the extraction or "harvesting stage" through manufacturing of finished products, use, reuse, recycling and disposal.

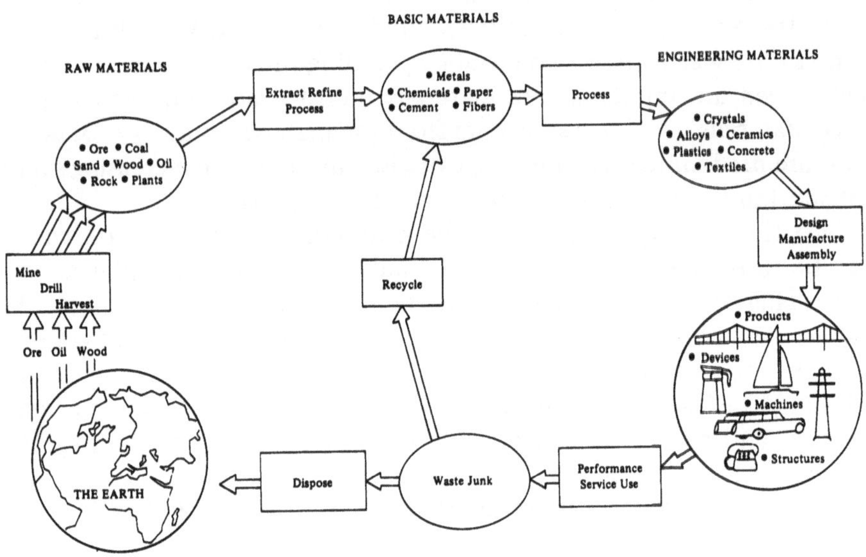

Figure 1: The Total Materials Cycle

Source: Materials & Man's Needs, Materials Science and Engineering, National Academy of Sciences Washington, D.C., 1974

The materials industries deal with the entire materials cycle. Looking at the many different stations of this cycle, it is immediately evident that only a multidisciplinary approach could deal with all the technological and market oriented questions related to the materials industry, where demands or pressures from the social and political arena are important inputs to provide appropriate solutions.

From the raw material to the finished product and its discard or recycling there are interactions with the following key elements:

– energy intensity of processes and products;
– environmental impacts;
– local labor and other social structures;
– innovation and technological leadership;
– free market forces versus national materials policy.

It is beyond the scope of this book to explain the complexity of the total materials industry that is interwoven into the majority of all segments of modern civilization and which is a key issue in the North-South Dialogue. In setting the stage for the book we have given sufficient indications that the total materials cycle is very much involved in the socio-political arena. Later, we will describe the status quo and perspectives of six key materials industries to arrive at a basis to explain why technology planning and technology assessment should be carried out.

3.2 The Problem Triangle

As Figure 2 indicates, the materials industry is subject to pressures (P) but also finds corresponding opportunities (O) within a certain triangle of forces. Furthermore, it should be kept in mind that the triangle actually "floats" on socio-political undercurrents.

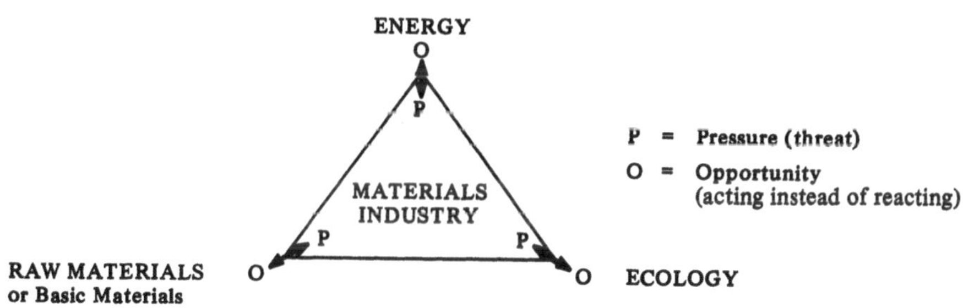

Figure 2: Problem Triangle of the Materials Industry

It is easy to see that the three forces: ecology, energy and raw materials, in many cases, should be handled as one integrated system, because none of them can really be completely separated from the others. Obviously, energy production requires fossil materials and minerals (nuclear energy). On the other hand, energy is not only an important element but often a significant cost factor in extracting materials from ores.

First, let us look in some detail at the ecology corner of the triangle in Figure 2. A pollution problem obviously exists when primary basic materials are produced. In fact, pollution results from misplaced, mismanaged or wasted materials. Materials can become "insults to the environment" at all points in the materials cycle: in mining and refining of ores, processing of materials into products, and in the disposal of wastes by consumers. Pollution is also an important issue in production of energy, regardless of whether energy is produced from fossil fuels or nuclear sources.

Therefore, technological solutions often have to deal with all three corners of this triangle and the technologies for survival of basic materials industries are often to be found within the triangle. This is true also for material-intensive manufacturing industries, like the car industry, which are involved with all three corners of the problem triangle.

The year 1972 was an important date for world-wide awareness of pollution problems. The Stockholm conference on environmental issues took place that year. It was presided over by Morris Strong, who afterwards, became the Director of the United Nations Environmental Program (UNEP). At Stockholm many developing countries declared that they were not willing to invest in pollution prevention for their new industries; arguing that basic materials industries, such as the steel industry, had polluted for decades. As a result the highly developed countries could carry out their successful industrialization process at rather low cost. Therefore, the developing countries would not agree to be pushed right away into a more sophisticated technology where environmental controls required extra investment for new production units.

But, after only 5 years, many developing countries followed suggestions from UNEP and other sources and decided to protect their environment while setting up new capacities to exploit their natural resources. Typical examples can be found in numerous countries in Latin America, Eastern Europe, Africa and Southeast Asia.

Therefore, a comprehensive understanding to keep emissions, from basic materials industry plants, at a tolerable level has spread within a few years time around the world. Of course, emissions must be collected in order to be abated. Often an abatement efficiency of the collected emissions of around 90% is considered appropriate, because "pollution zero" is idealistic but cannot be paid for in realistic terms.

Let us now turn our attention in more detail to the energy corner of the triangle. No other important issue of our time is so shrouded in fog as energy. Some of the mass media tell us that our future energy will come from renewable sources like geo, bio, wind and solar. But the hard truth is that, for the rest of this century, the sum of all these alternative energy sources can only provide a small fraction of the total primary energy needs of the world. For instance within countries in the Organization for Economic Cooperation and Development (OECD) solar energy, the front runner of the alternative energies (according to recent studies by the International

Institute for Applied Systems Analysis, Vienna) can supply a maximum of approximately 5% of primary energy by the year 2000. The main reasons are: high capital requirements, long lead times and a lack of proven technologies. The lion's share of solar energy in the next 20 years will be for heating and cooling of buildings plus provision of heat for agricultural or industrial needs. By the way, we do not count the use of new biomass for energy production as solar energy even though biomass contains stored solar energy. In the USA often all renewable energies combined are called solar energy. This includes hydropower and biomass and, could reach 20% of total primary energy use in the USA or around 50% in Brazil by the year 2000. All the other previously quoted alternative energies, can only provide a negligible amount of the energy in the industrialized countries.

In the developing countries the situation is somewhat different, because biomass is already a large contributor to the total primary input. Solar energy in selected sun-rich parts of the world may have an interesting share for their total primary energy input within the next two decades.

But for the production of basic materials, "soft" or alternative energies will be applicable only in a few cases. For the primary production stage, the majority of the basic materials will at least for the next three decades, rely almost entirely – as up to now – on energy inputs from fossil, nuclear and hydroelectric sources.

This will have specific importance in the future, when we look to the need to use ores with a lower metal content. The amount of energy needed to process these ores may even be prohibitively high. We are faced with this situation because mankind, in its use of the nonrenewable fossil fuels and minerals, has focused mainly on exploiting the more highly concentrated deposits – which are rather rare in the earth's crust. Gold is one example: the time when gold nuggets could be found at certain locations is over. Now more dispersed ores have to be separated from huge quantities of rocks. Let us consider for example today's mining of copper ores in North America. Here the energy aspect immediately enters consideration. To use leaner ores more energy per unit of output is needed and as energy becomes more expensive and shorter in supply there will be a lower limit at which extraction of metals from certain ores will not be economic. This again points out the strong interaction within the problem triangle.

Of particular importance for our book will be the third corner of the problem triangle, where raw materials and basic materials are placed together. Although they represent subsequent steps, their interrelationship with pollution and energy is rather similar. Furthermore, in this part of the triangle the free market forces come fully into play. The thousands of materials which now are used to keep the materials industry and consumer product industry going, are subjected to countless negotiations, price considerations and performance criteria.

In each industry, engineering department or purchasing directors finally decide within a specific cost-benefit-risk analysis what kind of material is, at this moment or in the foreseeable future, appropriate for their application or production.

There are problems in this third corner. National and international policies and regulations are difficult to achieve. Many objectives of the North-South Dialogue are unresolved.

Since the COSMAT study was published in 1974, great efforts have been made especially in the USA to establish a national materials policy. Also, endeavors are continuing to make sure that universities and other scientific institutions, as well as the government and public, realize the interaction of materials, energy and ecology, because there is a finite world.

A few leading universities in the USA have opened institutes or departments for "policy studies" for investigations on energy and materials. Europe lags in this respect. In any case, as of today, a systems approach to the problem triangle shown in Figure 2 remains a rare exception.

3.3 "The Limits to Growth" Syndrome

After the Club of Rome published their first study in 1972, numerous investigations looked into the question of future availability of fossil fuels, minerals and ores for the production of energy or materials. Today, there is high probability that the supply of fossil fuels, such as oil and gas, from presently explored sources in North America, the Middle East and Europe will dwindle in a few decades. But, as stated earlier in our book, we deal with non-fuel engineering materials and their corresponding raw material inputs.

The various studies on "limits-to-growth" caused considerable concern in industrialized countries as to whether their materials industry could count on a sufficient supply of raw materials for the foreseeable future. The discussion about this question has reached mass media all over the world and some premature conclusions soon emerged:

- generally a supply "pessimism" gained momentum regarding many imported raw materials;
- therefore, forecasting of "Zero Growth" for the materials industries in industrialized countries emerged;
- developing countries, rich in mineral resources, came to some mistaken conclusions like they would soon have a stronger political position and/or much higher income from exports of their minerals;
- most difficult of all were the efforts to establish multilateral treaties including buffer stocks and complicated mechanisms to regulate supply and demand of non-fuel raw materials.

The studies on the general subject of "The Limits to Growth" created a widespread fear that the world would, by lack of natural resources, soon run out of basic materials. This is in no way supported by facts for those primary sources which provide the lion's share of input for the materials industry.

In order to understand this, one must be aware of the difference between "reserves" and "resources"*. Reserves normally consist of developed mines plus known deposits which can be exploited at or near present prices with existing technology. The fact that some reserves are sufficient to last only 20–30 years should not cause any worry. This is a normal situation for many industries. Reserves are only meant to assure production for about 20 years and provide a buffer to allow time for discovery and development of new reserves.

Resources, on the other hand, consist of:

1) subeconomic deposits which will become reserves when and if prices rise to a level which makes their development economically attractive or when technology sufficiently lowers production costs;
2) known deposits which, for one reason or another, are unlikely to be developed within the planning time frame;
3) undiscovered deposits.

The preceding can be illustrated with the aid of the diagram shown in Figure 3.

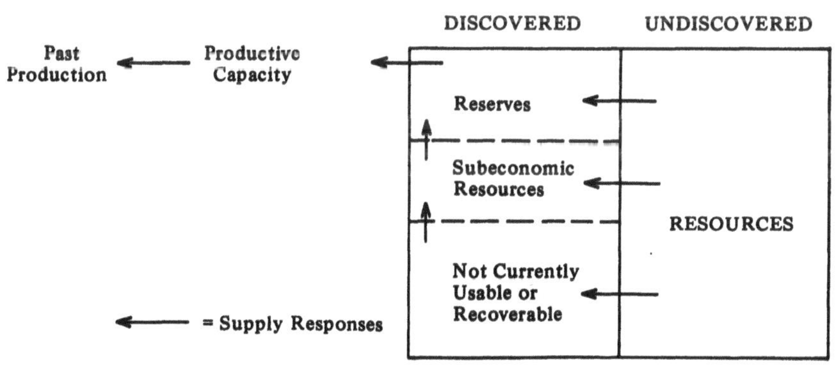

Figure 3: Dynamics of Reserves and Resources

* The US Bureau of Mines and the US Geological Survey (Geological Survey Bulletin 1450-A, 1976) now classify "reserves" as a part of "resources" with many sub-categories for both designations. Reserves are considered "identified resources" and these are further broken down into "demonstrated", "measured", "indicated" and "inferred". However, since these definitions differ from those used in other countries we have chosen to use the terms which distinguish between "reserves" and "resources".

It is instructive to begin with the right half of the diagram. Since knowledge of the location and size of the ore bodies within the earth's crust is incomplete, once a deposit is discovered it can be placed into any one of the three categories shown on the left-hand side of the diagram. The classification depends upon current and expected future prices and the state of technology. A newly discovered deposit may be so rich that it is immediately classified as "reserves" or, it may have clear development potential, but only at somewhat higher prices, and thus be classified as "subeconomic". Finally, a discovery may not be sufficiently promising other than to take note of its existence and general location, since it is so marginal relative to existing reserves that is does not appear economically exploitable within the foreseeable future. It would be put into the "not currently usable" category.

If demand is high and new discoveries of high-grade deposits are not large enough to fill the gap, the price rises, thereby shifting some "subeconomic" deposits into the category of "reserves". Such a shift into reserves would also occur as new technology reduces the cost of mining, beneficiation, smelting, and/or transportation of raw or semifinished materials. Furthermore, as technological changes accumulate, as major technological breakthroughs occur, and/or as prices rise substantially, deposits classified as "not currently usable" can move upward into the "subeconomic" category. Usually such shifts occur slowly.

Another important aspect is the rate at which reserves are transformed into "past production" which in turn depends upon current production and demand. Past production constitutes a resource for recycling which is another reason for not worrying about resource depletion. Most common metals are eligible for recycling up to 50% or even more.

Already today, gold and silver are carefully recycled. Around 40% of the copper and lead are recycled, and aluminium is at the 28% level. The recycling of steel products, such as used cars will gain enormous momentum in the next decade or two. Scrap may contribute up to 50% of the steel industry output by the turn of the century.

The authors of "The Limits to Growth" asked the right questions like "Are there limits to the earth's supply of resources? " However, their computer models were misinterpreted. In Chapter II we will show that there is no reason to believe that the world is running out of minerals.

The main argument against "Limits to Growth" is the "Growth of Limits". This applies not only to growth of reserves but also to the use of different "ores", for example, nodules from the sea floor. In addition, there is already today in industrialized countries a drastic shift towards less-materials-intensive technologies and products.

Since all evidence indicates that there will be no physical lack of resources for the next century and probably for many generations thereafter, let us look at factors which could affect the availability and supply of minerals, ores and basic materials.

1) Political-geographical constraints: partly, these are subjects of the North-South Dialogue. However, since so many different countries are eager to supply ores for common metals, severe constraints by cartels are rather unlikely. An exception (temporary) to this may be the supply of chromium, cobalt and manganese mainly from regional sources: southern Africa and the USSR. These three metals are used as alloying elements by the steel industry and are being stockpiled by consuming countries like the USA and West Germany. However, in the medium term, substitution will solve any scarcity problem. For instance, stainless steel containing up to 10% chromium can be substituted by other steel alloys, titanium or aluminium depending on the application. Furthermore, according to a recent study by the Office of Technology Assessment in the USA, most of the common metals can be substituted, up to 50%, rather easily by other metals or non-metallic materials like plastics.
2) The time lag to provide new capacity for the production of ores, concentrates, primary metals: there could be shortages during the eighties because incentive is lacking to invest in mines or corresponding basic industries and/or to maintain their function.
3) Import restrictions by consuming countries because of balance of trade deficits.
4) Massive price increases for minerals and metals because of more costly energy.
5) Population growth.

The factors enumerated above will in the future require much closer monitoring of situations and trends. In the past, markets served as the indicators which generated the signals for the pattern of resource development. Now uncertainty about compliance with environmental legislation, longer lead times and greater capital requirements for plant construction together with an increasing interdependence among the nations of the world, will require a more sophisticated mechanism for evaluation, assessment and planning.

3.4 Resulting Trends in OECD Countries

If we look at OECD countries in three geographical areas (USA, Japan, Western Europe) the differences are striking: the USA imports only 10% of their materials and 20–25% of their primary energy, whereas Japan is almost totally dependent on imports to cover its energy and materials needs. Western Europe is in a middle position, importing about 80% of its needed minerals and 60% of its primary energy. Let us for a moment concentrate on the situation of the Western European industrialized OECD countries.

Although there will be no minerals resource constraint for basic raw materials for many decades, temporary and severe constraints are quite possible from the political arena regarding access to, and pricing of, the raw materials needed by Western Europe. Furthermore, the production of many basic materials is energy intensive and therefore sensitive to curtailment or price increases for primary energy. For these reasons the capacity of the basic materials industries like steel, primary aluminium or basic chemicals will hardly be increased in Western Europe and it is even doubtful whether older obsolete plants will be replaced. Instead, Europeans will tend to seek cooperative agreements to set up new production units in places where natural resources, like energy and raw materials, are in abundance. But when looking at the issue of the necessary employment of people in the Western European materials industry, it is obvious that growth resulting from innovation and new technologies is the best "escape route". Up to now the dominating criterion for success was market pull which, however, is a rather short-lived phenomenon. On the other hand, Europe (like Japan) is confronted with the necessity for strategic investment decisions which are not initiated by market pull but by a long-range perspective for 10 or 20 years ahead. In Western Europe and Japan this has first priority, because they could be confronted with resource problems already within one decade. Therefore, government and industry should favor investments which reduce the amount of imported raw materials by suitable innovations, because only through innovation will industries remain competitive. But, simultaneously social innovation is necessary to achieve appropriate growth by using less materials per unit of Gross National Product (GNP) than in the past. This means to turn away from "planned obsolescence" or the "throw-away" society which was dominant in many industries until the mid-70's.

Up to that time, the consumption rate of many primary materials grew more rapidly than the increase in GNP. The coupling factor ("elasticity") was above 1 for steel, approximately 2 for aluminium and above 2 for plastics. From now on, in the industrialized countries, we will see an accelerated "uncoupling" when growth rates of GNP and materials consumption are compared.

This uncoupling will be achieved through materials conservation which will take the form of recycling and reuse of materials, miniaturization, and designing products with a longer lifetime. In addition, the growth of the service sector and its contribution to GNP will also result in less primary materials input per unit of GNP. But nevertheless, world-wide consumption of primary materials continues to increase further each year. Thus, it becomes necessary to search for progressive perspectives of future basic material supply. Large materials industries in industrialized countries cannot afford to wait in the frame of North-South Dialogues for multilateral agreements between the different governments. In this context it would be most desirable that large companies in OECD countries establish bilateral agreements with developing countries rich in resources or even trilateral agreements. The latter could comprise an industrialized country as the source of advanced technology, a developing country as the resource partner and, an OPEC development fund or a consortium of banks as the financier. Such an agreement would establish a new basic materials industry close to the resources of common interest. Multinational or transnational corporations are one of the possible bridges to enhance new investments in developing countries under a fair deal scheme.

There is no long-range solution for the problems in Western Europe and Japan without Third World cooperation. Third World countries have a demand for products from OECD countries and they also have resources for the materials industry. To cope with the problems of employment and supply security in Europe and Japan, suitable solutions with the Third World must be found and the materials industry has a specifically high importance in this matter.

3.5 Materials Consumption and Economic Growth

As stated earlier, the highest consumption of materials is in countries of the North Temperate Zone. For the most part this means the industrialized countries. Figure 4 illustrates that they have gone through a prolonged period of rather steady growth in per capita consumption of materials.

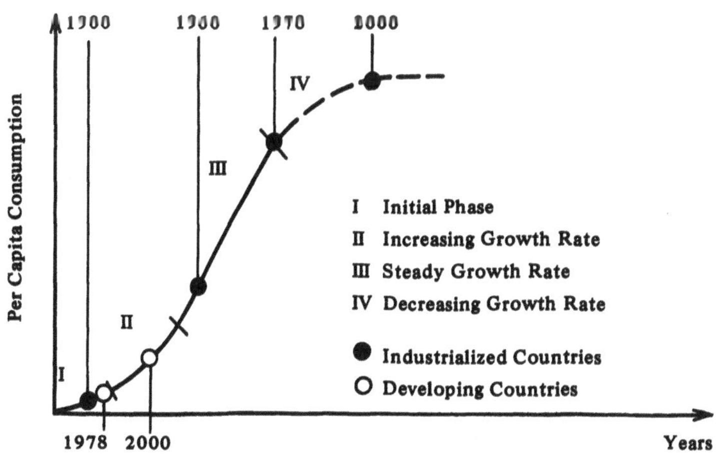

Figure 4: Development of Per Capita Raw Material Consumption (Schematic)

In most of the industrialized countries for instance between 1950 and 1974, the per capita consumption of steel increased annually by 4–6% and that of aluminium by 8–10% to just quote two examples. This brings up the questions "When will developed countries enter Stage IV of the curve or are they already in it?" and "When will per capita consumption in the industrialized countries level off or start to drop?" These are the key issues. For many phenomena, the change in slope of the curve usually occurs because of internal dynamics, *not* because of external pressures. Why or when the inflection point occurs is difficult to analyze or to forecast.

In projections to the year 2000, the industrialized countries high share in materials consumption will continue, but will undergo percentage-wise decreases as developing countries and middle income countries grow in relative economic importance.

Growth of world GNP is desired and anticipated. But, what will the growth rates be? Can industrial raw material inputs keep up with expanding rates and higher levels of GNP? In order to try and come up with answers to such questions, more than one method of forecasting has been proposed. The applied yardsticks include such terms as "elasticities" and "intensities of use".

The elasticity for a material in a country, is computed by dividing the growth rate of consumption for the material by the growth rate of GNP. Intensity of use (IU) is obtained by dividing the actual tonnage usage of a material in a given country, by that country's GNP.

There is a general relationship between demand for mineral resources and GNP. In developing regions, materials consumption tends to increase more rapidly than GNP. In Africa for example, the growth rate of aluminium demand has remained relatively constant at around 12% per year for more than twenty years and during the same time period GNP has been increasing at around 3% resulting in an "elasticity" of around 4. In other words, aluminium consumption is increasing at a rate four times that of GNP in Africa.

Although the method of calculating elasticities seems simple, a problem arises when either one of the growth rates becomes negative. Such a situation is not covered by the definition of elasticity. Another problem arises when growth rates are fractional because large elasticity numbers can result. Both problems arose, in industrialized countries, in recent years. As a consequence, year-to-year comparisons of elasticity cannot be interpreted. However, in general, elasticities for materials consumption are expected to decline for the rest of this century.

In the industrialized countries, the increasing relative importance of the contribution by services to their GNP reduces the contribution by materials. This means that less materials input is required for each unit of GNP. Figure 5 illustrates this for the USA. There, it can be clearly seen that the

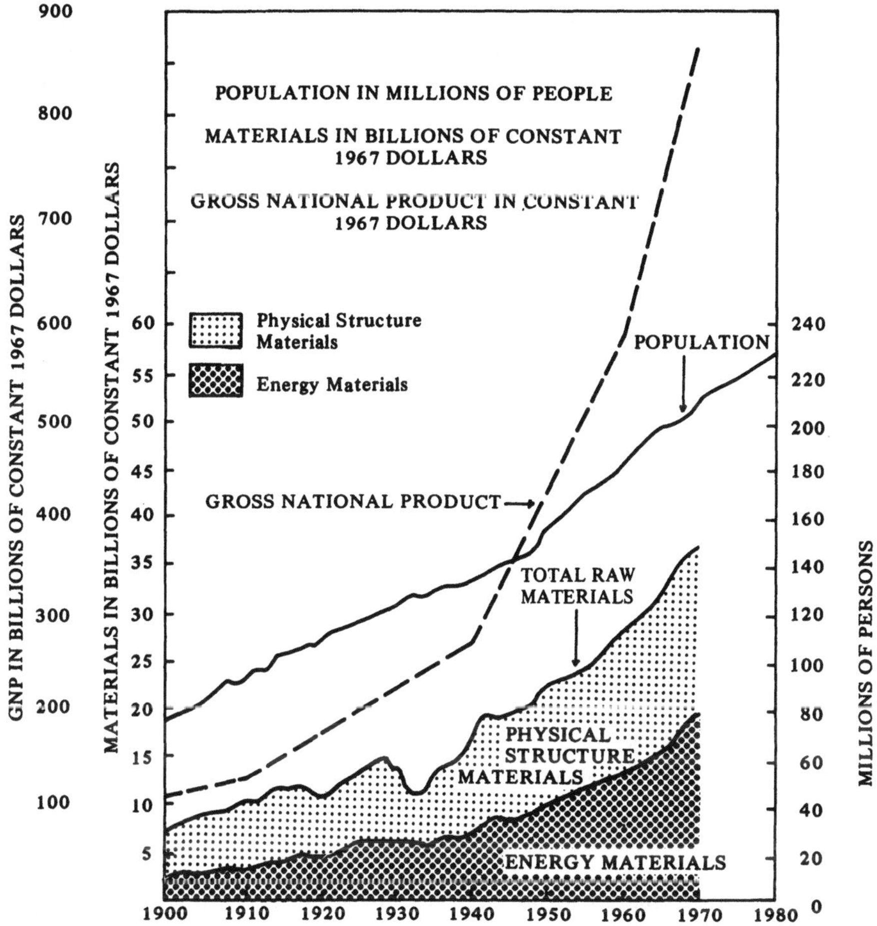

Figure 5: Raw Materials in the United States Economy 1900–1969

Source: Bureau of Census, US Department of Commerce; and Bureau of Mines, US Department of the Interior.

rate of materials consumption is slowing in relation to GNP growth. Until 1973, the value of total raw materials consumed in the USA was less than 5% of the GNP and non-energy materials represented about half that value.*

It has been argued that the intensity of use (IU) which is obtained by dividing the actual usage of a material in a given country, by that country's GNP, explains such a phenomenon better. When the value thus obtained is plotted versus GNP per capita, a curve results as shown in Figure 6 for steel. This figure illustrates that the intensity of use, for steel, has decreas-

* Further details can be found on pages 169–170.

ed in 16 industrialized countries. In most of these countries, consumption of common materials and metals is growing slower than GNP, but notable exceptions to this are aluminium, plastics and platinum which today and in the foreseeable future continue to grow faster than the GNP.

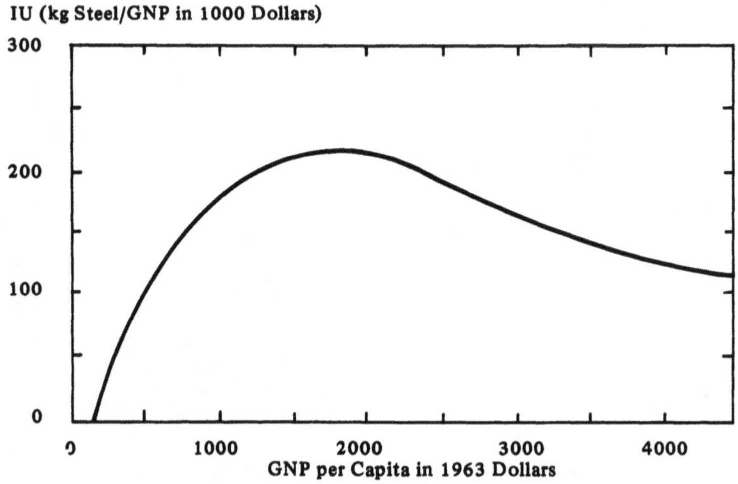

Figure 6: Intensity of Use for Steel in Sixteen OECD Countries
Source: Organisation of European Aluminium Smelters (OEA) 1976/77

However, declining elasticities and curves showing decreasing intensities of use do not mean that total demand for a given material is decreasing. For example, world demand for steel is expected to double by the year 2000. And, when one considers that world population growth rates are less than projected materials demand growth rates, this means an increase in per capita consumption. Therefore, although elasticities and IU's are declining, per capita consumption continues to grow for most materials albeit at diminishing rates.

All of this indicates a greater need for technology planning by the basic materials industries.

3.6 Shift of Basic Materials Industries to Developing Countries

As Table I shows, within one decade, production volume of basic materials made considerable progress in developing countries. But the lion's share remains till today in the developed market economies. However, in the next 2 decades, growth rates of basic materials production are expected to be considerably higher in less developed countries and centrally planned economies than in developed market economies in the Western World.

Table I: Changes in the Distribution and Growth of the Production of Selected Individual Processed Materials in the World Economy: 1960–1969*

		Percentage Share of World Production		
		Developing Countries	Developed Market Economies	Centrally Planned Economies
Metals:				
Crude Steel	1960	2.6	67.3	30.1
	1969	4.1	67.5	28.4
Aluminium	1960	1.8	81.3	16.9
	1969	8.3	77.7	14.0
Copper	1960	17.2	66.9	15.9
	1969	19.3	63.2	17.5
Ceramics:				
Cement	1960	13.5	59.8	26.7
	1969	16.5	57.0	26.5
Chemicals:				
Resins, Plastics	1960	0.2	91.0	8.8
	1969	1.3	88.7	10.0

* Derived from data in Industrial Development Survey Vol. V, United Nations Industrial Development Organization, United Nations, New York, 1973

Despite the expected high growth rates in the developing countries, these rates should be kept in perspective. It must be realized that they are starting from a low base, which means that the major portion of world manufacturing is likely to remain predominantly in the already industrialized countries in the visible future. Shifts and changes will proceed gradually. While these are taking place, the industrialized nations can adjust their manufacturing sectors towards high-technology, high-productivity industries to replace the more basic industries which would have greater economic advantage in developing countries.

The independence of many resource-rich previous colonies after World War II created the strong trend to industrialize these countries and to create added value in the countries owning natural resources in demand. But during the quarter century between 1950–1975 many mistakes were made by introducing unappropriate technology to developing countries. The "unappropriateness" resulted from choosing large scale and/or sophisticated production plants too early, creating so-called "cathedrals in the desert".

In the early 70's, an intense North-South Dialogue began to arrive at new concepts of interdependence. At the same time the less developed countries, especially those rich in resources, asked for new models of cooperation of various kinds, i.e., to set up buffer stocks for stabilizing prices as well as suitable treaties. Another key issue is a "code of conduct" governing technology transfer to these countries. In addition, they wish aid to help them establish end uses for their materials, in their countries, by application of appropriate technologies. The full basket of requests will take many more years of negotiations which will lead to bilateral or multilateral treaties.

Recently, there has been technology transfer between developing countries. This South-South cooperation is called Technological Cooperation between Developing Countries (TCDC) in United Nations language. Under these circumstances, the "technology donor" will often be an "industrializing" developing country, like Brazil, Venezuela or South Korea. For the OECD countries and their materials industries this represents an opportunity more than a threat. A tripartite agreement could be arrived at whereby the partner from an industrialized country would play a "give-and-take" role in providing financing, feasibility studies and, marketing and/or taking surplus production.

The common denominator in these examples is easy to recognize: world changes in the materials industries should have, as a main desirable aim, a movement towards one another of the Third World countries and OECD countries, developing their resources in a common interest.

3.7 Concluding Remarks

So far, all this sounds logical and perhaps the too simple conclusion is that the industrialized countries and the developing countries have largely similar or complementary interests.

Even if we leave aside social and political pressures and complications and concentrate mainly on the technological and industrial aspects, including important price relationships of various materials, there remain plenty of problems which are poorly defined and not clearly recognized. However, these can be solved with available or modified (appropriate) technology, or new technologies within reach in a decade or two; provided that planning is carried out properly and that systematic, detailed understanding can be brought about.

Countries at all stages of development are involved and each of them can expect important benefits from a better understanding of the opportunities and problems related to basic materials. Therefore, the main task is first to recognize and define such opportunities and problems and then to indicate appropriate solutions from the viewpoint of industries dealing with the total materials cycle world-wide.

In the next chapter the outlook for selected industries is discussed.

Bibliography

Lester R. *Brown*, "World without Borders", Vintage Books, A Division of Random House, New York, August 1973

Economic Report of the President, US Government Printing Office, Washington, D.C., January 1977

D. *Gabor* et al, "Beyond the Age of Waste", Pergamon Press Ltd., Oxford, New York, 1978

Herman *Kahn* et al, "The Next 200 Years", William Morrow and Company Inc., New York, 1976

Herman *Kahn*, "World Economic Development 1979 and Beyond", Westview Press Inc., Boulder, CO, 1979

Wassily *Leontief*, Ann P. Carter, Peter A. Petri, "The Future of the World Economy", Oxford University Press, New York, 1977

Library of Congress, Science Policy Research Division, Congressional Research Service, "Materials Policy Handbook: An Outline of Legislative Issues of Materials Research and Technological Application", Congress Serial C., US Government Printing Office, Washington, D.C., 1977

D.H. Meadows et al, "The Limits to Growth", Universal Books, New York, 1972

Dennis L. *Meadows* and Donella H. Meadows, "Toward Global Equilibrium: Collected Papers", Wright-Allen Press, Inc., Cambridge, Massachusetts, 1973

National Academy of Sciences, Committee on the Survey of Materials Science and Engineering (COSMAT), "Materials and Man's Needs: Materials Science and Engineering", National Academy of Sciences, Washington, D.C., 1977

National Academy of Sciences, Committee on Mineral Resources and the Environment (COMRATE), "Mineral Resources and the Environment", Washington, D.C., National Academy of Sciences, 1975

National Commission on Materials Policy, "Materials Needs and the Environment Today and Tomorrow", US Government Printing Office, Washington, D.C., 1973

National Commission on Supplies and Shortages, "Government and the Nations Resources", US Government Printing Office, 1976

Organisation of European Aluminium Smelters (OEA), 76–77, May 1977

Unido, Industrial Development Survey, Vol. V, United Nations, New York, 1973

US Bureau of Mines, "Mineral Trends and Forecasts", Washington, D.C., 1979

II Present Structure and Future Trends in Key Materials Industries

1 INTRODUCTION

Before we can approach the main purpose of this book, to explain and interpret changes for the materials industries, first we need to review the present structure of key materials industries, because numerous readers may not be familiar with the status quo of these industries.

The introductory chapter dealt, in more general terms, with the question "Is there a resource constraint for the key materials industries?" The first part of the present chapter will give the actual resource situation and an overview for the most important industrial metals. In the following sub-chapters, the structure and future trends of six key materials industries are outlined, briefly covering their materials flow from feed stock to final uses. Current and future technologies are also discussed.

In concluding this chapter, we ask the question "Where are the basic materials industries heading?" This indicates problems and opportunities, furnishing reasons to examine new mechanisms to deal with the current and future situation of the basic materials industries. Thus providing a "spring board" to explain the methodology of technology planning and technology assessment as a means to obtain an analysis and to arrive at an adequate data base.

1.1 Overview of 5 Major Metals and Their Reserves

The five major industrial metals are steel, aluminium, copper, zinc and lead. Iron and steel, copper, and aluminium will be dealt with at greater length in separate sections later in this chapter. They are only given a brief overview here.

Table II provides data on 4 non-ferrous metals and their proven ore reserves for the decade 1966–1975. Comparing the first and the fourth column in this table we can see that within a decade the proven reserves of aluminium (bauxite) have been tripled, zinc ores increased by factor of 2.5, copper ores 2.3 and lead about doubled due to more intensive exploration efforts. These increases took place despite considerable continuous mining of these ores during this decade. Comparing the second and third column, for copper, in Table II is a real "eye-opener": five times more new copper reserves were proven as were mined during this decade.

Table II: Proven Ore Reserve Development for Copper, Aluminium, Lead and Zinc in the World (in million tons)

Metal (ore)	Proven Reserves on 1.1.1966	Consumption	New Proven Reserves		Proven Reserves on 1.1.1976
		in the years 1966 to 1975			
Copper[1]	195	– 63	+ 324	=	456
Aluminium (Bauxite)	5 964	– 605	+ 11 913	=	17 272
Lead[1]	93	– 33	+ 115	=	175
Zinc[1]	75	– 54	+ 164	=	185

[1] Metal content

Source: United States Bureau of Mines, Washington, D.C., and Bundesanstalt für Geowissenschaften und Rohstoffe, Hannover

In the case of bauxite the situation is even more favorable: 20 times more bauxite was found than used! Within this decade, consumption amounted to 600 million tons and at the same time almost 12 billion tons of new bauxite deposits were proven. The total proven reserves of bauxite would be sufficient to last about 300 years at 1978's rate of consumption.*

If we were to look at the reserves of iron ore, according to the US Bureau of Mines, the 1978 consumption rate of iron and steel can be maintained at least 200 additional years by using the known deposits of iron ore without counting the huge amounts of probable resources.

Therefore, we can say for the common metals, there is no reason to believe that within the foreseeable future, or even within one century, industrial production would be endangered because there would be a shortage of minerals. Even in a purely hypothetical case that one of the metals might be in short supply, for example copper or manganese, this would never be a situation compared to a catastrophic development. Rather, the material would become more expensive and its price would

* It is interesting that in "Government and The Nations Resource", December 1976, a report of the National Commission on Supplies and Shortages in the .USA, the known world reserves of aluminium are estimated to last only 23 years. This proves again that there is plenty of misleading information around, partly triggered by the self fulfilling prophecies of studies like "The Limits to Growth".

continuously increase over a period of years. Therefore, other materials would take its place. In fact, this has happened to the consumption of tin where its price today is about 14 000 dollars per metric ton – 10 times more expensive than copper! The reason for this is not a depletion of tin ore but the price policy of the four producing countries which have caused not only this price increase, but great concern to all users of tin who have to depend on this artificially controlled metal. Further, copper production has not changed considerably in the seventies. This can be seen in Table III which compares production of 7 non-ferrous metals since 1950. Of course 1975 was a "black" year for almost all metals due to the recession.

Table III: Western World Production of 7 Non-ferrous Metals (1000 mt)

Year	Aluminium	Copper	Zinc	Lead	Nickel	Tin	Magnesium
1950	1 288	2 860	1 810	1 678	118	175	17
1955	2 591	3 323	2 301	2 047	195	175	79
1960	3 618	4 198	2 438	2 141	264	156	67
1965	5 095	5 044	3 128	2 428	303	153	125
1970	8 056	6 163	3 951	3 056	455	187	169
1971	8 621	5 837	3 764	2 915	463	189	180
1972	9 205	6 371	4 093	3 013	427	197	178
1973	10 129	6 682	4 247	3 091	498	189	182
1974	11 096	6 923	4 350	3 104	555	185	193
1975	9 898	6 272	3 750	2 869	535	183	172
1976	10 223	6 635	4 124	3 042	562	185	161
1977	11 314	6 855	4 231	3 126	525	184	182

Production Change in Five Year Intervals (%)

1950/1955	+101.2	+16.2	+27.1	+22.0	+65.3	0	+364.7
1955/1960	+ 39.6	+26.3	+ 6.0	+ 4.6	+35.4	–10.9	– 15.2
1960/1965	+ 40.8	+20.2	+28.3	+13.4	+14.8	– 1.9	+ 86.6
1965/1970	+ 58.1	+22.2	+26.3	+25.9	+50.2	+22.9	+ 35.2
1970/1975	+ 22.9	+ 1.8	– 5.1	– 6.1	+17.6	– 2.1	+ 1.8

Source: Metallgesellschaft AG, Frankfurt a.M.

It must be kept in mind that estimates or projections of metal reserves as well as demand or utilization rates can both be in error. Therefore, the cumulative effect of such errors could give rise to either unjustified pessimism or optimism. On the one hand, we are optimistic that even for a nominal growth of the materials industries there are enough resources available through the next century.

On the other hand, although our planet has large resources, we must not forget that there are several other growth constraints such as: energy, pollution and social limits to growth.

If one considers that population is increasing further, even a 2% increase in per capita consumption can bring about many problems.

Today, almost 90% of man-made engineering materials are used in the industrialized countries. Let us therefore ask the question, "Where are the resources located?"

More important than the question of world-wide scarcity of resources is the regional distribution of the reserves. In most cases the countries in which there are reserves are not those which are the centers of consumption. In some cases there is also a very high regional concentration of reserves. Therefore the question of access to the raw materials could be of much more importance than the overall physical availability.

As a rough description of the regional distribution of the reserves of most metals, it can be said that 40% of them are held by the industrialized countries, 30% by the Eastern countries and another 30% by developing countries. Of the reserves to be found in the industrialized countries, more than 80% are in the USA, Canada, Australia and South Africa. Russia has more than 80% of the reserves of the COMECON countries. Among the developing countries the predominant share of reserves is held by a very limited number of countries, mainland China being the front runner. Consequently, not only Western Europe and Japan, but also most of the East European countries and about 70% of all developing countries have only very limited reserves of minerals. Therefore, we should not overlook problems which could arise in arriving at new agreements within the North-South Dialogue between resource-rich developing countries and regions who need to import basic materials.

1.2 Profiles of 12 Important Metals

In this book it is not possible to provide in-depth detail on all metals. However, we have selected 12 important commercial metals for a profile analysis.

They are:	Cadmium	Magnesium	Tin
	Chromium	Manganese	Titanium
	Cobalt	Molybdenum	Vanadium
	Lead	Nickel	Zinc

Of the five major metals which were previously given an overview, only lead and zinc are included here.

An analysis of the regional concentration of the reserves of specific metals is quite informative. It is interesting in this context to identify those minerals for which there is both a high regional concentration of reserves and an extremely high dependence of industrialized countries on supplies. Table IV gives a condensed summary of countries supplying the majority of important ores.

Table IV: Regional Distribution of Selected Raw Materials

	Raw Material	Share of 3 Countries 1977	Share of 5 Countries 1977	Regional Distribution of Measured and Indicated Reserves-1977 Country and Percentage Share
BASIC METALS	**Iron**	59.4	76.7	USSR (30.2), Brazil (17.5), Canada (11.7), Australia (11.5), India (5.8)
	Copper	44.9	58.7	USA (18.5), Chile (18.5), USSR (7.9), Peru (7.0), Canada (6.8), Zambia (6.4)
	Lead	47.8	61.4	USA (20.8), Australia (13.8), USSR (13.2), Canada (9.5), South Africa (4.1)
	Tin	50.2	68.1	Indonesia (23.6), China (14.8), Thailand (11.8), Bolivia (9.7), Malaysia (8.2), USSR (6.1), Brazil (5.9)
	Zinc	45.8	58.6	Canada (18.7), USA (14.5), Australia (12.6), USSR (7.3), Ireland (5.5)
LIGHT METALS	**Aluminium**	62.8	74.8	Guinea (33.9), Australia (18.6), Brazil (10.3), Jamaica (6.2), India (5.8), Guyana (4.1), Cameroon (4.1)
	Titanium	59.0	74.1	Brazil (26.3), India (17.5), Canada (15.2), South Africa (8.6), Australia (6.6), Norway (6.4), USA (6.0)
ALLOYING METALS	**Chromium**	96.9	97.9	South Africa (74.1), Zimbabwe Rhodesia (22.2), USSR (0.6), Finland (0.6), India (0.4), Brazil (0.3), Madagascar (0.3)

	Raw Material	Share of 3 Countries 1977	Share of 5 Countries 1977	Regional Distribution of Measured and Indicated Reserves-1977 Country and Percentage Share
ALLOYING METALS	**Cobalt**	63.0	83.5	Zaire (30.3), New Caledonia (18.8), USSR (13.9), Philippines (12.8), Zambia (7.7), Cuba (7.3)
	Manganese	90.5	97.7	South Africa (45.0), USSR (37.5), Australia (8.0), Gabon (5.0), Brazil (2.2)
	Molybdenum	74.3	86.9	USA (38.4), Chile (27.8), Canada (8.1), USSR (6.6), China (6.0)
	Nickel	54.5	76.8	New Caledonia (25.0), Canada (16.0), USSR (13.5), Indonesia (13.0), Australia (9.3), Philippines (9.0)
	Vanadium	94.9	97.2	USSR (74.8), South Africa (18.7), Chile (1.4), Australia (1.4), Venezuela (0.9), India (0.9)

Source: "Facing the Future", OECD, Paris, 1979

For four metals, (chromium, manganese, molybdenum and vanadium) three-quarters or more of the measured and indicated reserves are found in only three countries. The most striking examples seem to be chromium, manganese and vanadium in which South Africa and the USSR predominate.

Now, let us look at the profiles for the 12 metals.

Cadmium: There is no resource constraint for this metal. Cadmium is obtained as a by-product from production of other metals (mainly zinc). It is used primarily as a protective plating on iron and steel. Cadmium stearate is used as a plastic stabilizer and there are many cadmium-based pigments. It is also used in nickel-cadmium batteries.

Chromium: More than 90% of the reserves of chromium are situated in only two countries – South Africa and Zimbabwe Rhodesia. Chromium is mainly used for metallurgical, chemical and refractory purposes. There is no substitute for chromium in stainless steels and no good substitute in hard plating applications. It is an essential alloying constitutent in high temperature alloys and is a rather critical material for almost all industrialized countries.

Cobalt: This metal is used principally in heat and corrosion-resistant alloys and in permanent magnets where it is an essential component. It also serves as a binder material in tungsten and other carbide cutting tools and in hardfacing alloys. There is no satisfactory substitute for cobalt as a binder in the hard carbide tool materials.

Zaire is one of the main suppliers of cobalt. Political and military turmoil within the country during 1978 brought cobalt production to a virtual standstill. If the situation continues it could have a drastic effect on certain industries.

Lead: The main application for lead is in storage batteries for automobiles, other transportation vehicles, portable power devices and emergency power units. The metal is used in gasoline anti-knock additives – but this use is declining due to poisoning of catalytic convertors. Because of its toxicity, lead has been replaced in interior paints by titanium and zinc pigments. A very high percentage of the lead used is being recycled because of its use in batteries which provides a ready source of material. Production of primary lead will most likely stagnate at today's level and in the next century go down.

Magnesium: There are almost unlimited resources for magnesium in sea water and brines. It has been called a "sleeping beauty" among metals and could find increased usage in automotive applications due to concern about energy. Because of its low density its use in die-cast parts could help reduce the weight of automobiles. However, usage is extremely price sensitive. But, if the magnesium to aluminium price ratio, on the basis of weight, were to fall below 1.25:1, magnesium castings could replace aluminium and other metals in the automotive market on a large scale. Today most magnesium is consumed in refractories and chemicals; only a small portion is used in metallic form. Magnesium metal is also used as a reducing agent in the production of titanium, vanadium, zirconium, uranium, and beryllium and as a deoxidizer in producing various alloys.

Manganese: This metal is essential in the production of steel, 7 to 9 kg required to deoxidize and desulfurize each ton of steel. No adequate substitute has been found in spite of intensive research. Manganese is also an alloying element for steel and copper alloys and is used in the production of aluminium and magnesium alloys. "Mining" of manganese (ocean) nodules could greatly increase availability.

Molybdenum: The United States has about 40% of the world's reserves of this metal. Molybdenum's major application is as an alloying element in: high-strength and stainless steels, high temperature alloys, cutting tools, hardfacing alloys, and magnets.

Nickel: More than 90% is used in the form of metal and alloys. It imparts strength and toughness to steel, and it is an ingredient of stainless steels, high temperature alloys, and copper-base alloys; coinage is a growing application. Other uses include electroplating, catalysts, batteries and fuel cells, hardfacing alloys, ceramics (bonding metal to enamel and glass), and in hydrogenation of fats and oils. New Caledonia and Canada have about 40% of the world's reserves of nickel. Ocean nodule mining is a possible future source.

Tin: Approximately one-third of this metal is used in the manufacture of cans and containers. Other major applications include bearings, bronze alloys, and solders. Aluminium and plastics are replacing significant amounts of tin in cans. While tin cannot be completely replaced in solders, its content can be reduced, with lead and antimony substituting for some of the tin. Increasing substitution and a considerable increase in price (due to cartel action) have kept its consumption fairly constant over the past ten years (see Table III).

Titanium: Titanium is the ninth most abundant element in the earth's crust. Because of its relatively high cost, titanium usage is presently limited to aerospace and chemical processing applications; current production processes use considerable amounts of energy, and relatively little metal is recycled. Present applications include jet engines, airframes, chemical processing vessels, and piping for corrosive fluids. If production costs could be significantly reduced, the metal could find increased application. Actually it could be another "sleeping" metal with a great future after the turn of the century.

Vanadium: Its main use is as an alloying element in steel, titanium alloys, and some high temperature alloys. Vanadium is also used as an oxidation catalyst and in the manufacture of welding rods. Russia has almost 75% of the world's known reserves.

Zinc: The construction industry is a major consumer of zinc – mostly in the form of galvanized steel, where no adequate substitute has been found for zinc. Another large amount goes to the transportation industry. Aluminium die castings and molded plastics are increasingly replacing zinc die castings. Canada, the USA and Australia have more than 45% of the world's known reserves.

1.3 Africa as a Regional Raw Material Source

In the preceding section we have touched upon regional sources of raw materials. Now we would like to comment in some detail on Africa as an example. The continent is estimated to have over 30% of the world's mineral resources.

Africa is a major supplier of cobalt, vanadium, chromium and manganese ores, providing 1/4 up to 1/3 of the world's production. The chromium supply, within the Western World, comes almost completely from the southern part of Africa.

Africa's copper production amounts to 20% of the world's production. Africa's share in world production of other basic materials like bauxite, iron ore, nickel, tin, zinc and lead is between 5 and 10%.

However, it is important to know that Africa's reserves of minerals are percentage-wise considerably higher than its current production. For instance, over 30% of the bauxite reserves are in Africa, but production is only about 10% of the world total. The same is true for a number of other raw materials for example manganese ores, where Africa has about 1/2 of all known reserves.

Another important point is that Africa's share in the world production of minerals has decreased in the last decade. The reasons for this are obvious: a recent survey of twelve mining companies revealed that they spent 51% of their exploratory money in Africa in 1961, but in the year 1975, less than 1%*. This will change only when the investment of private capital is safeguarded by international treaties. Establishing a new mine requires enormous investments in fixed capital. And, when one considers that the life of a mine is at least 20 years, this means that the investor must have confidence that his investment is protected for the life of the mine. Today, this is a very important aspect in resource availability.

1.4 Disruption in Mineral Supplies

As implied in the previous sections, the regional location of many ores, under certain circumstances, could lead to a supply interruption. But, is there a way to cope with such a situation? Is it possible to have advanced warning of impending interruptions?

As a step towards a national minerals policy and an aid for long-range planning, the Interior Department of the USA is developing an "indicator" of potential disruptions in the supply of "critical minerals". The development of the program received added impetus after internal warfare in Zaire caused cobalt shortages and a 700% price rise.

Although world-wide physical resources of many minerals appear to be sufficient to meet projected demand for generations to come, short-term supply disruptions with their associated rapid price increases, can pose a significant threat to industry. Therefore, the indicators will include a "disruption index" that would rank minerals according to the likeli-

* Source: Metallgesellschaft AG, Frankfurt a.M.

hood of such market disruptions as producer-cartel developments, sudden drop-offs in exportation and significant abrupt price boosts, and a "cost index" that would measure the economic implications to the overall United States economy should the market disruptions occur.

The US Interior Department decided to develop the indexes for raw materials supplies such as cobalt, bauxite, copper, nickel and iron ore initially, and then expand them to include lead, zinc, manganese and others. The indicators will replace the existing over-simplistic approach of labeling certain minerals as critical materials.

So far, the Interior Department's survey shows that there is a high probability of continued market disruptions in cobalt supplies through 1990.

The primary benefit of the indicator would be the creation of an "early warning system" having great economic importance to mining and metals producing industries.

Now, let us turn our attention to the structure and future trends in some selected industries.

2 IRON AND STEEL*

Iron has been used by man for at least 5000 years. It became well known and widespread during the "Iron Age", which started about 1200 B.C., and left its impact on ancient civilizations.

More than 95% of iron production is converted into steel, the rest is processed mostly into cast iron products. Steel is an iron alloy, containing a low carbon content, small amounts of alloying elements and controlled impurities. It has improved properties over those of cast iron.

Steel has contributed to modern civilization more than any other metal. With its introduction on an industrial scale, 120years ago, the steel industry played an important role in the industrial revolution.

Steel production started to gain momentum after World War II. From 1945–1975, it increased world-wide from 100 million tons to nearly 700 million tons per year, showing an average annual growth of nearly 6%. During the period 1975–1978, steel which was always a symbol of economic power, became a symbol of crisis.

The declining demand for steel in the Western World resulted in low utilization of capacity. Fierce competition, among most steel producers and exporters, especially by Japan, Taiwan, Korea, and the Eastern bloc countries, accompanied by price erosion, deteriorating profits, plant closures and uncertainties regarding the future, caused deep concern to steelmakers in the USA and Western Europe.

In 1979 the economic outlook for the steel industry improved. On examining the present status of the steel industry, the following facts may be reported:

The world's crude or raw steel production capacity is estimated at 900 million tons per year. The world's production was 713 million tons in 1978, utilizing approximately 79% of available capacity.

The Western World's steel production capacity is estimated to be around 650 million tons per year. With production of approximately 468 million tons in the Western World (1978), the capacity utilization was only 72%.

The expectation that by the year 1980 world steel production might reach 1000 million tons has been abandoned. The growth of steel consumption in important industrialized countries like the USA and Western Europe has been around 2% in 1978, less than the growth of GNP, which was around 4% in the USA and 2.5–3% in Western Europe.

Despite the fact that, in 1978/79, there was a considerable overcapacity for steel production in Western Europe, there are further ongoing investments in developing countries to set up new steel producing facilities.

* In cooperation with I. Reznik, Dipl. Ing., Swiss Aluminium Ltd.

They may be divided into three categories:

– *Conventional plants using modern blast furnace technology* and medium-sized capacity of at least 1–3 million tons per year.

– *Direct reduction process plants,* mostly based on a local supply of natural gas and having modular production units, each with a capacity in the range of 400 000 tons per year. They are located for instance in several Middle East countries, Venezuela, Brazil, Canada, and USSR (Kursk).

– *"Mini plants"* which have a capacity of around 100 000 up to 400 000 tons per year, or even less (like the "Bresciani" plants in Italy). Electric remelt furnaces are the main equipment, using steel scrap as feed stock.

2.1 Consumption and Perspectives

Steel is a relatively young metal. Within over one hundred years it showed tremendous growth, mainly in the field of construction, transportation and machinery.

Various types of steel were developed, such as tool steels characterized by high hardness and resistance to abrasion, stainless steels which resist corrosion and high strength, low alloy steels (HSLA). The HSLA steels are made by combining the effects of minor concentrations of alloying elements with the control of rolling and quenching operations in the steel mill. The cost of these steels are only minimally higher than low-carbon steels. The HSLA steels are being used more and more where higher strength-to-weight ratios are an advantage, for example, in automobiles. The weight of these steels is 10–30% less, to perform a given function in a car, than the weight of the materials replaced. Their importance to the automotive industry is expected to increase (see p. 174 for a few more words on this innovative move by the steel industry). Steel, which is used more than any other metal, has reached by now a certain maturity in technology and in its application. The evolution of its consumption and growth are shown in Table V.

The consumption of steel is related to the development activity in each country, and therefore has an interesting behavior pattern (Figure 6, page 16). It has been established, that growth of steel consumption increases up to a certain degree of economic maturity (GNP per capita). Then, the growth of steel consumption slows down in relation to a further rise in the GNP per capita.

The reason is that during development and industrialization, GNP growth is largely a function of a high share of capital formation in the

Table V: World's Steel Consumption 1900–2000

Years	Total World Consumption in Million Tons per Decade	Growth per Decade in %	
1900–10	520	–	increase of growth
1910–20	700	+ 35	
1920–30	900	+ 29	
1930–40	1200	+ 33	
1940–50	1700	+ 42	
1950–60	2700	+ 59	
1960–70	4800	+ 78	
1970–80	6900*	+ 44	decrease of growth
1980–90	10200*	+ 48	
1990–2000	13800*	+ 35	

* Estimates

Source: The World of Metals, by M.F. Dowding, Metals and Materials, July 1978

building of industrial plants and infrastructure, all of which is highly steel intensive; this results in a growth rate of steel consumption which exceeds the growth rate of GNP. At later stages, when the main steel-intensive infrastructure (railroads, other transportation etc.), has been built, the importance of steel in relation to the GNP declines.

The 1976 crude steel production and consumption per capita in various countries and regions is shown in Table VI.

Table VI: Crude Steel Production and Consumption per Capita (1976)

	Production kg / capita	Consumption kg / capita
Sweden	632	740
USA	507	620
USSR	565	570
Japan	940	540
EEC	516	450
Latin America	84	92
China	31	35
Africa	20	20
India	15	18

The consumption in representative industrial countries is in the range of 450–750 kg/cap., whereas in developing countries it amounts to 20–100 kg/cap.

In industrialized countries, iron and steel are used mostly in transportation, construction and machinery. In the USA, for example, these three sectors cover nearly 80% of the total steel consumption. Transportation uses 32% of total steel consumption, and is the main sector of application (road, rail, sea-transport). Construction uses 26% of total steel consumption. In machinery, 20% of the total steel consumption is used. There are numerous other uses such as in canning and containers, in the gas and oil industry, high-voltage electrical transmission towers, consumer goods etc.

What are the perspectives until the year 2000?

Steel will maintain its leading position among common metals. Its world production in 2000 might reach 1.2–1.5 billion tons*. In other words, within about 20 years, the world production of steel is expected to double. A change in market share has already started to take place. The developing countries, are expected to double their share of production, from a projected 12% in 1980 to at least 24% in 2000. The Eastern bloc will roughly preserve its share (from 28% to 26%). The share of the developed world will decrease from 60% to 50%.

2.2 Geographical Distribution

Iron ore is used for the production of crude iron ("pig iron" or "hot metal"), the first step in the conventional steelmaking process. Iron ores are available in many parts of the world. They are mined in over 50 countries, being used either domestically, or exported.

In 1976, the total world's ore production was over 890 million tons (iron content of more than 500 million tons). The major ore producing countries were: USSR (28% of world's total), Australia (11%), USA (10%), and Brazil (9%). Other important producers are Canada (6%), France (5%), India (5%), Liberia (4%), and Sweden (3.5%).

Approximately 400 million tons of ore (nearly 50% of total ore production) are exported. The main ore exporting countries in 1976 were:
Australia (21% of world's export), Brazil (17%), USSR (13%), Canada (11%), Liberia (9%) and, to a lesser extent, India (5.6%), and Venezuela (5%).

The main ore importing countries were Japan (36% of world's import), West Germany (12.5%) and USA (12%).

* Estimates by US Bureau of Mines, and M.F. Dowding, respectively.

PRODUCTION
1978: 713.0 million metric tons

"Other" consists of:

China and DPR Korea	4.8%
Other Asia	2.8%
Latin America	3.4%
Africa and Middle East	1.5%
Oceania	1.1%

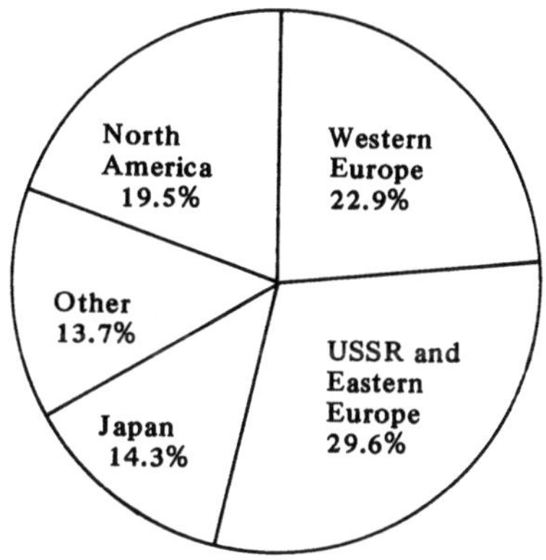

CONSUMPTION
1978: 713.0 million metric tons

"Other" consists of:

China and DPR Korea	6.3%
Other Asia	4.5%
Latin America	4.2%
Africa and Middle East	4.2%
Oceania	0.8%
Unallocated	– 1.0%

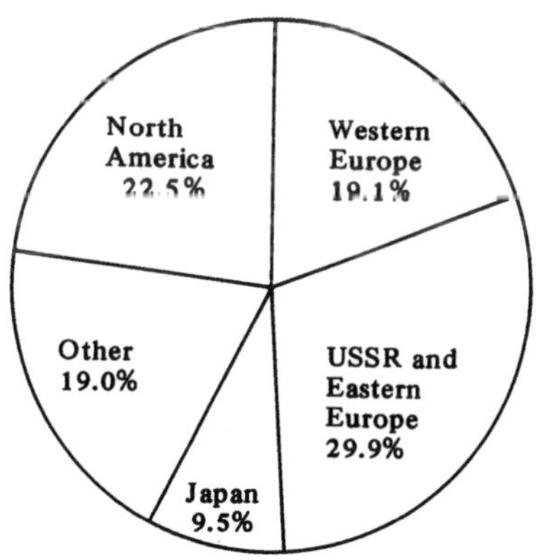

Figure 7: Geographical Distribution of Steel Production and Consumption 1978
Source: IISI/World Steel in Figures / 1979 Edition

A huge sea and land transportation network serves the iron ore trade. The main sea routes are: to Japan from Australia, India and Brazil; to Europe from West Africa, and, to the USA from Venezuela.

The geographical distribution of steel production and consumption in 1978 is given in Figure 7.

2.3 Production of Iron and Steel

Pig iron is produced from iron ores in blast furnaces. The dominant part of it is further processed into steel. Steel is conventionally produced from pig iron and, iron and steel scrap by various processes, as shown in Figure 8.

"Sponge iron" produced by direct reduction accounts today only for a minute fraction of world steel production. It is shown therefore by dotted lines.

Most of the produced steel is cast into ingots or continuously cast into blooms, billets or slabs, for further rolling and finishing operations. A minor portion goes to steel foundries.

Raw Materials

Iron ore: According to the US Bureau of Mines, the world proven reserves of iron, in ores, are approximately 90 billion tons. An additional 100 billion tons of iron in ores might be available from other sources, mainly in North and South America.

The largest proven reserves are found in the USSR (30% of world's total), Brazil (18%), Canada (12%) and Australia (12%).

Smaller reserves are found in India (6%), USA (4%), and in various other countries, like Liberia and Sweden.

Ores with high iron content are found in Brazil and Liberia (68% iron content), in Australia, Venezuela, Mauritania (65%), and in Sweden (63%). Ores with medium iron content (50–60%) are found in North and South America, the USSR, India and South Africa.

Low-grade ores with 35–50% iron content are found in Eastern and Western Europe where, except for France and Sweden, ore production is now very small.

The cumulative world demand of iron in ore from 1976 to 2000 is estimated at approximately 18 billion tons (of iron content). Today's proven world reserves are already 5 times more. Assuming that additional reserves will be found by the year 2000, no shortage in iron ores is foreseen even for the next whole century.

The problem facing the steel industry is therefore definitely not a physical shortage of ores. However, the fact that there is too little capital

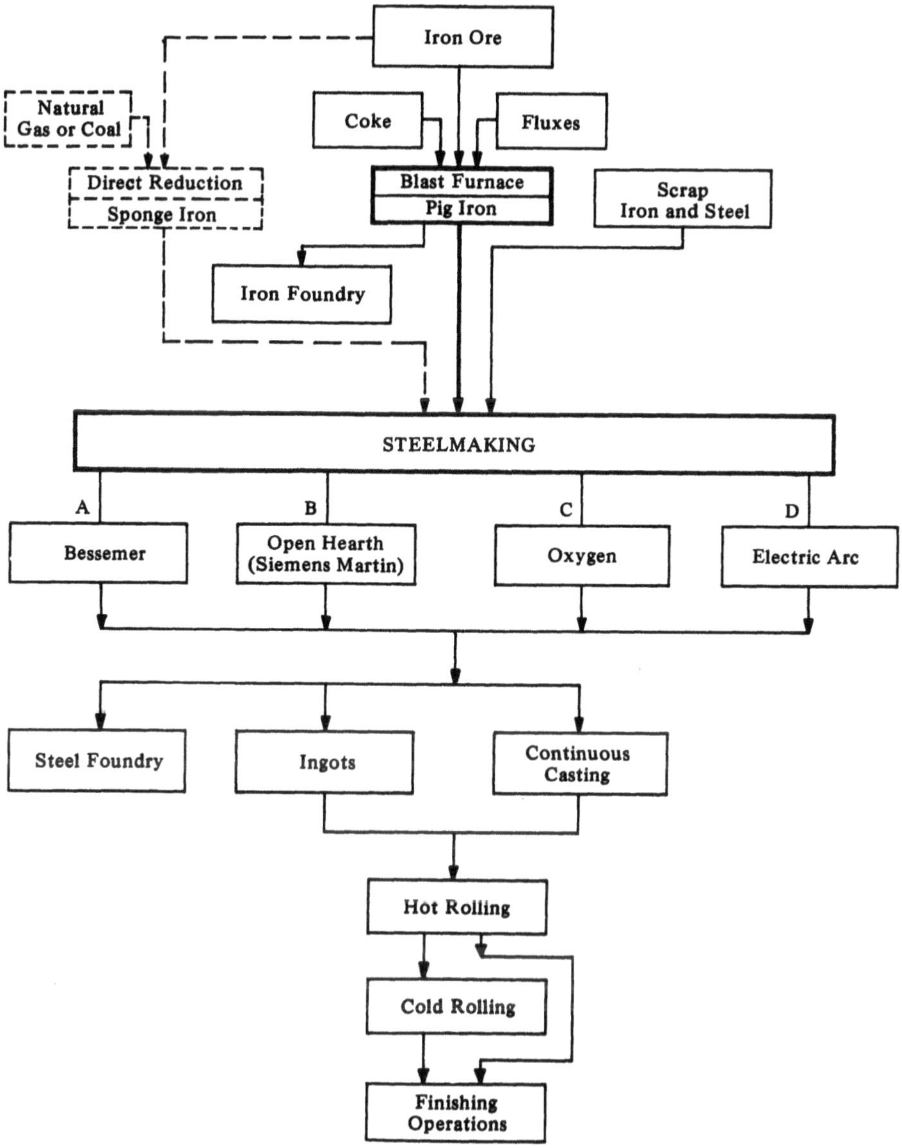

Figure 8: Iron and Steelmaking Processes

investment in ore mining (due to poor profitability of the steel industry) might lead to a temporary supply shortage in the next decade or so.

Scrap: Iron and steel scrap are important input materials in steelmaking and in foundry production.

In the USA, in steelmaking, approximately 45% scrap and 55% pig iron are being used. The proportion of iron and steel scrap, and pig iron, varies according to the type of furnaces used in the steelmaking process. Steel and iron foundries use up to 80 or 90% scrap.

The consumption of iron and steel scrap in 1977* was in the USA 84 million tons, in Japan 35 million tons, and in West Germany 20 million tons. The scrap portion in steel production will probably increase in the future, as shown in Figure 9, but this forecast is not unchallenged.

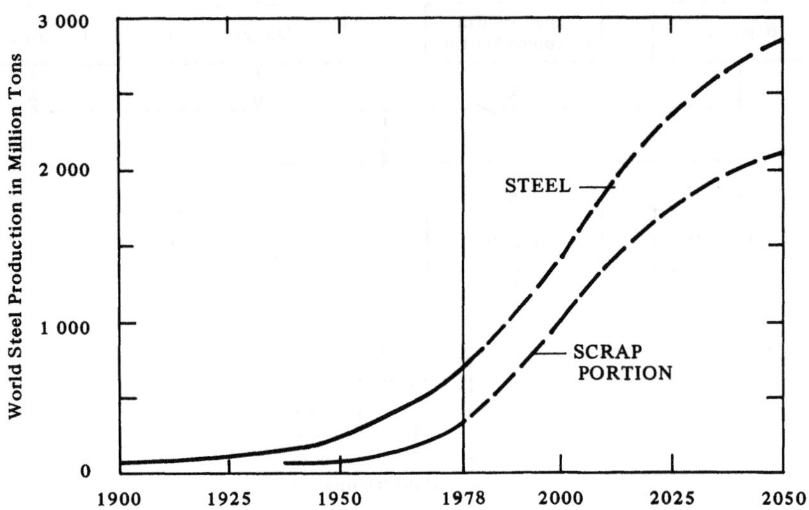

Figure 9: Scrap Portion in Steel Production

Source: After W. Dettmering

Increasing scrap shortages and volatile prices are main concerns of the steel industry world-wide.

The trend to operate "mini-mills" (small steel production units), which require nearly 100% scrap, international trade barriers and scrap trade deficits – e.g., in Japan, Italy etc. – will have their effect on scrap shortage and price.

* World Steel Figures, IISI 1979 Edition

More intensive scrap recycling, and in parallel, lifting international trade barriers, could probably help in solving the above problems.

Coal: For the steel industry, metallurgical coke (produced from "coking coal") is an important material input. It is used in blast furnaces, where "pig iron" is produced from its ores. In the Western industrialized countries coke consumption is 440–600 kg per ton of pig iron produced.

There are 3 types of steelmaking countries, those with:
- plenty of coking coal: USA, USSR, Poland, China, West Germany, Australia;
- no coking coal but gas: Saudi Arabia, Algeria, Nigeria, Iran, Venezuela;
- neither: Japan, Brazil, Egypt.

Large coal resources exist in the USA (over 50% of the world's coking coal resources), Western Canada and Australia as well as in China and the Eastern bloc countries. Although coal reserves are plentiful, potential shortages and expected price increases are a world-wide problem. The supply of coking coal remains therefore one of the important problems of the steel industry. The utilization of lower grade coals which are abundant in nature, might help to alleviate the situation.

2.4 Beneficiation and Transport of Iron Ores

Iron ores are mostly surface mined. Underground mining is seldom used due to higher mining costs.

Owing to the demand for higher iron content and uniform chemical composition, almost all iron ores are beneficiated before shipping. High-grade ores are processed mainly to improve their size uniformity; low-grade ores, to increase their iron content, by concentration methods.

After crushing and screening, the ores consisting of small size particles (up to 6 mm) are agglomerated into "sintered" particles or into "pellets". This is done to salvage the fine ores, improve blast furnace performance, and to facilitate transportation and mechanical handling. Pellets and sintered ore are widely used today. In the USA, for example, over 50% of the iron ores consumed were in pellet form.

Ore transportation overseas and inland became a huge operation. Ships of over 200 000 tons capacity transport ores for instance from Brazil to Japan. Several oil/ore bulk-carriers of nearly 300 000 deadweight tons, are being used for ocean transport in recent years.

Transport of iron ore slurry by pipeline is increasing. Such pipelines are already in operation in Brazil, Mexico and Argentina (the "Marcona-process" is used to a large extent in slurry transport).

2.5 Utilization of Ores for the Production of Iron

Ironmaking: After suitable preparation, as concentrating, sintering or pelletizing, iron ores – or partly reduced ores – are converted into "pig iron" mostly by using the conventional blast furnace.

Recent developments of "direct reduction" processes permit production of "sponge iron", which can replace pig iron or scrap in producing steel or foundry products.

Several such "direct reduction" processes were introduced whereby high-grade ore is reduced preferably with natural gas.

Direct Reduction Process

Direct reduction is a method of producing "sponge iron" directly from ore without going through the molten stage as occurs in the blast furnace. In blast furnaces the hot metal temperature is as high as 1400°C whereas in direct reduction the maximum temperatures obtained are less than 1100°C. The direct reduction process involves the intimate contact of prepared iron ore with a "reductant". The reductant can be a gas or a solid. When natural gas is used, it is first heated and "reformed" to a mixture of carbon monoxide and hydrogen gases and then passed at a controlled temperature, into the furnace, where iron oxide is reduced to iron. However, only 1% of world steel production resulted from direct reduction in 1978.

Iron Ore Outlook

According to the US Bureau of Mines the world demand for iron ore is expected to double from 500 million tons iron content in 1976, to 1000 million tons in the year 2000. The main use will remain as feed stock for blast furnaces.

Direct reduction, from ore into metal, will be further developed and will be mostly utilized by countries with extensive gas resources and no coking coal of their own.

Radical changes in iron ore production technology are unlikely. However, improvements in technology to increase mining and beneficiation efficiency are already underway.

2.6 Iron and Steel Processing

Principally, conventional steelmaking is a two-step process. First, iron oxide is reduced in blast furnaces, with coke, to pig iron. Second, the pig iron is refined in steelmaking furnaces through oxidation processes which reduce mainly the carbon content and remove impurities.

Pig iron is produced from iron ore in blast furnaces. There, the ore containing iron oxides and impurities is reduced to molten crude metal. The furnace is charged with iron ore, sinter or pellets, coke, and fluxes, which assist in the above process. Pig iron is tapped in a molten state and transported for further processing, mostly into crude steel, or cast into ingot molds.

Crude steel is made from pig iron and, iron and steel scrap. It differs from pig iron by its lower carbon content, which makes it resilient, flexible, stronger and more workable than pig iron.

The main steelmaking processes and the subsequent working of steel are described further in 2.7.

2.6.1 Investment and Production Costs

The iron and steel industry is capital intensive. The investment required for operating a modern mine in the USA, designed to produce 3–10 million tons of iron ore products per year, is between $ 150–600 million.

The investment required for a medium size, integrated steel plant of 3 million tons per year would cost $ 3 billion. The production costs of steel vary from one plant to another. The main cost components are given in Table VII.

Table VII: Cost Structure of Finished Steel Products (USA)

Cost Element	% of Total Cost
Ore	20
Energy (coal, fuel etc.)	20 – 25
Wages and Salaries	35 – 45
Capital Charges (incl. depreciation)	6 – 10

2.6.2 Energy

Since 1973, oil price more than quadrupled and, gas and coal prices went up considerably. In view of expected shortages and price increases of energy in the future, the steel industry faces some difficult problems.

Among the items listed in Table VII, energy is an important cost factor, as it accounts for 20–25% of the total production cost of finished steel products.

The main energy sources, in the USA, in 1972, when steel production was nearly at its peak (133 million tons) are shown in Table VIII.

Table VIII: Energy Used by the USA Steel Industry (1972)

Energy Sources	% of Total
Coal	63.0
Fuel Oil	6.0
Natural Gas	24.0
Purchased Power	4.1
Other	2.9
Total	100.0

The three fossil energies, listed in this table, serve a dual purpose: to provide the process heat and to reduce iron oxide into iron. Coal, converted into coke, is the main energy source, and will probably remain so for many years. Natural gas and fuel oil will surely become more expensive and scarce and will have to be replaced, at least partly, by energy derived from coal; and at a later stage, probably beyond the year 2000, by nuclear energy in combination with partly non-fossil reductants like hydrogen.

Such changes, to replace oil and gas by coal, involve a huge development in coal mining, transportation, and adaptation of new technologies. Obviously, it will impose a heavy financial burden on the steel industry and hamper its capability to meet market requirements. Some developments in the direction of energy conservation are already taking place. Efforts to optimize the energy balance of the blast furnace show promising results. Trials show that by various measures, e.g., injection of hydrocarbons (mainly oil) a drastic decrease in coke consumption could be achieved (from approximately 500 kgs coke per ton of pig iron, to approximately 250 kgs coke per ton). This is of course very important for countries in the Middle East and Africa, which have no coking coal but gas or oil.

The highest waste of energy in steel plants is in the coke making operation, where red-hot coke is quenched with water. This will be replaced in countries with energy (and pollution) constraints by dry quenching of coke with nitrogen. The heated nitrogen will be used for making steam or preheating the coal on its way into the coking chamber.

In order to "size" the importance of the energy factor in the steel industry, the industry's energy consumption in relation to national energy consumption is given in Table IX. It must be remembered that energy cost will continue to increase in every country and, therefore, will become a more important matter as time passes. The saving of energy or the increase in the efficiency of energy use are important factors in the iron and steel-

Table IX: Rate (%) of Energy Consumption of the Steel Industry to Total National Energy Consumption

	United States			Europe			Japan		
	1960	1970	1980	1960	1970	1980	1960	1970	1980
Energy for Steel Industry	6.0	5.2	4.4	12.4	10.3	8.7	16.2	20.7	19.1

Source: OECD Energy Commission

making process. Many steel works have established procedures in order to minimize waste or loss and reduce total energy consumption. The energy accounting of steel is described on page 147.

2.6.3 Environment and Pollution Control

Environmental problems and the cost of pollution control are topics of top priority in the iron and steel industry.

In mining and beneficiation, problems concern mainly disposal of solid waste, (e.g., tailings), elimination of dust, especially in crushing and grinding operations, reclamation of process water and, recultivation of mined out areas.

In iron and steelmaking, the industry is concerned, in many sectors, regarding pollution by solid waste, liquids and gases.

Solid waste includes slags from the iron and steelmaking furnaces, flue dust, and mill scale. Slags are sold and used for various purposes like additives to concrete etc. Flue dust and mill scale (mostly iron oxides) are used in sintering plants for recycling purposes.

Liquid waste includes water and various acids, oil, grease etc.

Air pollution by dust and fumes from steel plants is one of the most expensive environmental control problems, considering the size of the industry. Coke ovens are the source of gaseous pollutants (such as sulphur dioxide), and dust. Steel furnaces have a variety of emissions including: dust, CO_2 and fluor compounds from fluxes.

The investment in workers protection and mainly pollution control recently became an important cost factor. It is estimated that in the Western World it comprises 5–10% of the total production cost, and in the future, might add at least 15% to new capital expenditure. The investments required are enormous. For example, the steel industry in the USA has invested until 1977 approximately $ 4.5 billion, and another $ 8 to

$ 10 billion would probably be invested until 1983 for pollution abatement equipment.

2.6.4 Recycling

As already mentioned, scrap is an important source of material for the iron and steel industry, and should become more important in the future. Scrap can be divided into three main groups: home scrap, new (or industrial) scrap, and old scrap.

"Home scrap" (the scrap produced by steel plants or foundries) is recycled internally without leaving the plant area. This is a scrap of known quality.

Purchased scrap sources are "industrial", (rejects, trimmings etc.), and "old" or "obsolete" scrap, as often called by the steel industry. Obsolete scrap includes mostly finished products, such as old vehicles, rails and rail cars, all sorts of unused machinery, demolished ships, and other steel structures.

There are many specific problems in the recycling of steel from old scrap. For instance, a small copper content, coming from the wiring harness in obsolete cars, together with tin from remelting old tin cans, has a synergistic effect and gives steel made from such scrap undesirable properties.

2.7 Description of Steelmaking Processes and Finishing Operations

2.7.1. Definition of Different Types of Steel

In general steels can be divided into two main groups, according to their chemical composition, that is, 1) unalloyed steel and 2) alloyed steel. Both of these groups comprise: basic steels, high-quality steels, and special steels. Special steels, according to DIN and Euronorms, generally contain less than 0.03% sulphur and phosphorous each and often have a lower carbon content than ordinary steels.

2.7.2 Conventional Steelmaking Processes

In the following, steelmaking processes are described according to their historical appearance.

A. Bessemer Steelmaking: This process produces the general purpose "mild" (low carbon), steel. It uses molten pig iron as raw material. Air is blown through the melt to remove carbon and impurities.

The Bessemer process was introduced on industrial scale in 1865. In recent years it has been almost totally replaced mainly by the oxygen steelmaking processes which are more economical and less air polluting.

B. Open Hearth Steelmaking: This process, also known as the "Siemens Martin" process, produces all grades of carbon steel, including low alloyed steel, in normal batch sizes from 20 to 500 tons, although there are a few 900 ton furnaces in existence. It utilizes, in different ratios, molten or cold pig iron and steel scrap as feedstock; therefore this process has good flexibility. The charge is melted by burning a mixture of atomized oil or natural gas with air over a hearth. Carbon and impurities are removed by oxidizing the melt.

The process was introduced in 1875, and is still widely used, mainly in the USA and USSR. In Europe, because the process has low productivity, its use is rapidly decreasing.

Open hearth furnaces comprise 24% of the total world steelmaking capacity.

C. Oxygen Steelmaking: The process produces today almost all types of steel – from ordinary to high-quality, in tilting vessels up to 400 t capacity. It uses molten pig iron and scrap as feed stock. Sufficiently pure oxygen is blown onto the surface or through the bath to remove carbon and impurities, and for heating the melt to pouring temperature.

The top blown process[1] was introduced first in Austria by the VOEST company in 1952, and became the most widely used process in the steelmaking industry, due to its technical and economic advantages. The capacity of all types of oxygen blown furnaces is about 55% of total world steelmaking capacity.

The bottom blown process[2] is a variation of the basic oxygen process. It was developed in 1968 at Maximilianshütte, West Germany.

In this process oxygen is blown from the bottom of the furnace through the melt. It is said to have lower construction cost, higher production rates and, good process flexibility. The total world production is estimated now to be over 30 million tons.

D. Electric Furnace Steelmaking: Electric furnace steelmaking was introduced in 1910 and is widely used today. There are in fact three different electric furnace applications in the steel industry:

[1] Usually called "BOP", for Basic Oxygen Process, or "LD" for Linz-Donawitz, Austria, where it was introduced on an industrial scale.

[2] The process is called "OBM" for Oxygen Bottom Maxhütte or "Q-BOP" for Quiet-Basic Oxygen Process.

– large size plants producing a wide variety of special steels, and more recently, also basic or bulk steels;
– mini-steel plants with simple equipment and product mix ("Bresciani");
– electric heated "converters" where liquid steel is refined.

The amount of steel produced in electric furnaces has undergone rapid growth. In 1960, it was only 10%; in 1978, it was already 20%; and, for the year 2000, estimates indicate a 30–40% share of production.

A few more words about the two main plant types in operation today.

Large size plants produce ordinary to high-grade and alloyed steels, such as stainless steels, tool steels, etc. The furnace size is now mostly between 150 to 300 t. The input material is steel scrap or pig iron and in some countries, sponge iron from direct reduction processes. In full-scale plants, the required heat is produced by an electric arc. Other plants use induction, resistance, electron beam or even plasma as a heat source for very special steels. In recent years, an entire new generation of electric arc furnaces emerged, providing high throughput. The increased amount of available scrap opened the opportunity to enter the market for common steels on an increased scale. In addition, electric arc furnaces make it easier to comply with environmental regulations, which is an important advantage over conventional processes.

Smaller size plants, called mini-steel plants, are being introduced for small outputs of 50 000 up to several 100 000 tons per year. The mini-steel plants use arc furnaces with a charge capacity in the range of 20–100 tons. Most of them use scrap, others use direct reduced iron as feed stock or blend the scrap to meet the requirements of quality steels. The product-range covers mostly simple rolled products, such as reinforcement bars, rods, profiles for windows and doors etc.

Such plants are in operation in many countries. In Italy they are located around Brescia. The word "Bresciani" is today a symbol for profitable mini-steel plants. It is estimated that the world capacity of the mini-steel plants is about 40 million tons per year, or 5% of the world total.

The percent of world raw (crude) steel production, according to processes used, is shown in Figure 10.

2.7.3 Special Steels

There are 20 alloying components used for special steels, up to ten are added simultaneously. Special steels represent a science and technology of considerable importance. There is a wide variety of end uses and alloying compositions for special steels. We will mention only the more important ones:

– stainless steel is the largest item among special steels; the typical alloy contains 18% chromium and 8% nickel;

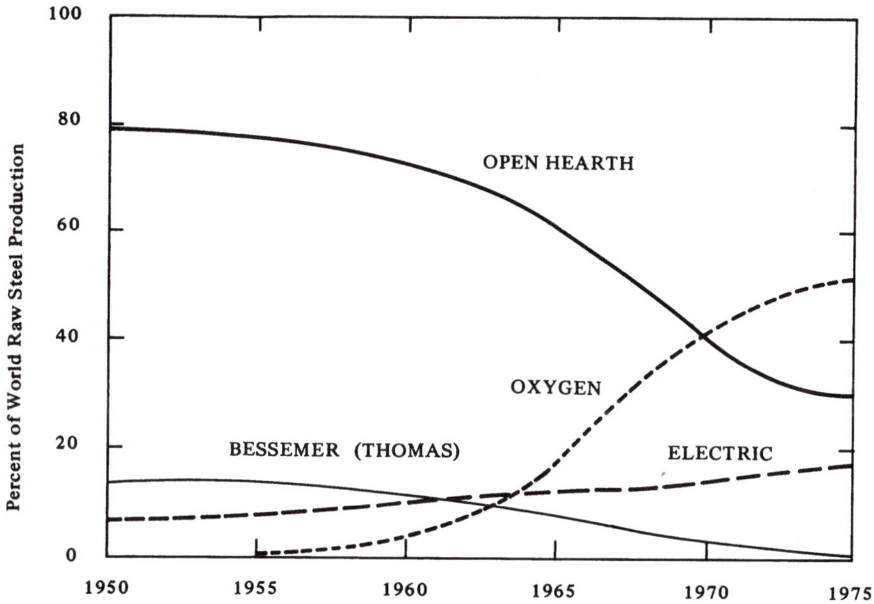

Figure 10: World Raw Steel Production by Process, 1950–1975
Source: US Bureau of Mines.

– corrosion resistant special steel may contain up to 30% chromium and is used for example in desalination units;
– heat resistant steels have chromium additions up to 10% and other alloying components;
– tool steels contain chromium and a wide variety of hardeners like cobalt, vanadium, manganese and others;
– iron-based or iron-containing magnetic materials contain a high amount of nickel.

To reach low impurity levels of sulphur, phosphorous and carbon, the liquid special steels undergo specific treatments.

A main goal in the production of stainless and other chromium-containing steels is the introduction of oxygen to the melt without oxidation of the chromium. To achieve this, the partial pressure of carbon monoxide within and above the melt must be reduced. This can be achieved by application of either vacuum and/or an inert gas like argon which is often combined with the introduction of oxygen into the melt.

About 20 years ago, vacuum treatment came into large-scale use and subsequently batches up to 70 tons were treated under vacuum.

In countries with surplus argon – resulting from air liquefaction units – decarburization was partly switched over to application of argon-oxygen mixtures in a specific converter (AOD-process). Right now, there is no longer an argon surplus and furthermore new, cheaper vacuum pumps make the vacuum treatment more attractive.

In conclusion: special steels offer, in basic and physical metallurgy, a fascinating field of high sophistication and further process development.

2.7.4 Finishing Operations

Semifinishing and finishing operations include casting of iron and steel, rolling, forging, coating etc. Nearly half of the capital invested and most of the labor costs are involved in these operations. Most of the processes are conventional and carried out according to common practice.

Casting: After tapping the molten steel from the furnace and following metallurgical treatments in the ladle, it is cast either into ingots or continuous strands.

A remarkable breakthrough in casting technique took place in Europe in the early fifties, when continuous casting was introduced. In this process, the steel is cast directly and continuously into strands, thus eliminating the previous need for mold casting of individual ingots. The process is characterized by high-yield and lower operating costs. Nearly 20% of Western World steel production is continuously cast. It will probably reach 50% by 1990. Certainly this is a most promising future development.

Hot Rolling and Forging: The solidified ingots are heated and then rolled in a primary mill into reduced sections by passing them several times through one or more sets of rolls. The products are known as semifinished steel. Reheated semifinished steel is further processed into flat rolled products (plates, sheet, strip, etc.) or into rails, structural or other shape, bars, wire rods, pipe or tubing.

Forging is done by hammering or pressing into the required shape. Normally a die is used. When one or only several pieces are to be forged (especially large pieces), "open die" forging can be used.

Cold Rolling: Cold rolling is used mostly to produce flat products. Cold rolled steel has better surface and dimensional characteristics than hot rolled steel, (approximately two-thirds of the flat steel in the USA is produced by cold rolling). In the process, material from a hot rolled coil is reduced in thickness, progressively, as it advances through several rolling stands. Most cold rolled sheet is given some form of heat treatment to restore ductility lost during cold rolling. Therefore, a typical cold rolling mill has heat treating and cleaning facilities for the processed steel.

Coatings for Steel: To prevent corrosion of the steel, several protective coatings were developed, mostly including metals such as: aluminium, chromium, nickel, zinc, tin, etc., as well as organic paints, varnishes, enamels, lacquers, and plastic. For flat rolled steel, the most important metallic coatings are zinc, tin, and chromium.

Galvanizing is a common coating process, where a prepared coil of rolled steel is passed through a molten zinc bath. Galvanized steel is used mostly in the automotive and construction industries.

Tin cans are made from tinplate, which is produced by a continuous electrolytic deposition process of a tin layer onto cold rolled sheet steel.

Tin-free steel, which is chromium coated steel, is used recently mainly for food cans and beverage containers.

Foundry: Iron and steel foundries, produce castings by pouring molten metal mostly into sand molds. Thus, almost any shape can be produced, whether full or hollow, simple or complicated. The trend today is towards larger foundries utilizing more mechanization and automation in molding preparation, pouring of castings and in subsequent handling.

2.8 Perspectives for New or Improved Technologies

Conventional Steel ("carbon steel")

There will probably be no revolutionary technological change in steelmaking by the year 2000, which could replace the current oxygen processes and electric furnace steelmaking. Future developments, as direct production of liquid steel from iron ore, are yet far away, and a complete continuous process from ore to finished steel product, is likely to remain out of reach for many years to come.

But experiments to produce liquid steel directly from iron ores (bypassing the blast furnace) and using a continuous oxidizing process, are being carried out.

For using sponge iron from direct reduction, a semi-continuous process is possible. An electric furnace is charged continuously with hot sponge iron. The tapped crude steel is cast continuously in strands, which are charged hot into the mill.

The use of heat, created by nuclear energy, to produce reducing gases from solid fuel, is being studied. The gases would be used to produce direct-reduced iron, which could be melted continuously in an electric furnace to produce crude steel.

The introduction of new technologies is time consuming and capital intensive. Therefore, it is very likely that the existing electric and oxygen steelmaking processes are here to stay, for at least the next two decades. But incremental improvements in process technology may take place in many segments of a steel plant. For example, continuous casting could have a breakthrough into continuous conversion of liquid steel into plates of approximately 20–30 mm thickness, rolled down immediately into coil, by tandem rolling mills. This development would be a big leap forward for the steel industry.

Other improvements are the use of cheaper oxygen with up to 40% nitrogen or use of hydrocarbon by-products in the blast furnaces.

New initiatives for basic technological improvements may well come from outside the steel industry like from the oil and chemical industries by introducing new feed stock for blast furnaces or from the aluminium industry, where strip casting is under rapid development. Further, the possibilities to save weight in vehicles by replacing steel with aluminium may well create an innovative spirit in the steel industry which could lead to the development of further improved medium or high strength steels (HSLA is perhaps just a beginning).

Special Steels

Here many new developments are under way. One trend is clearly evident, the use of two separate units one after the other:

1) an electric furnace to produce the liquid steel;
2) a converter for refining and alloying.

In the past, and on a large scale still today, both operations were carried out in the unit mentioned under 1). But, now the emerging trend is to upgrade 1) into a "melting machine" of high throughput with 2) providing the best geometric and thermal conditions for optimum refining and alloying. This technique is already being used for special steels and may well "spill over" into conventional steelmaking.

In the production of special steels, many different configurations for the application of vacuum, gases and alloying additions are in the production or trial stage, with the aim to reach improved combinations of cost and performance of the material.

The above remarks have referred to the production of special steels. When it comes to their alloy compositions, specific properties and end uses, the perspectives and ongoing innovations could fill an entire chapter. Suffice it to say that key issues, among others, will be magnetic properties, resistance against corrosion or high temperatures and, use in composite structures.

3 ALUMINIUM

Aluminium is a unique metal. There are many reasons for this statement. It is a "young" metal. Its commercial production did not begin until the second half of the 19th century. It was not until 1886 that an economical process of producing aluminium by electrolysis from a fused salt bath was discovered by two men who worked independently, Charles Martin Hall, an American and Paul L.T. Héroult, a Frenchman. In the relatively short time since then, primary aluminium production has undergone an amazing growth (Figure 11)).

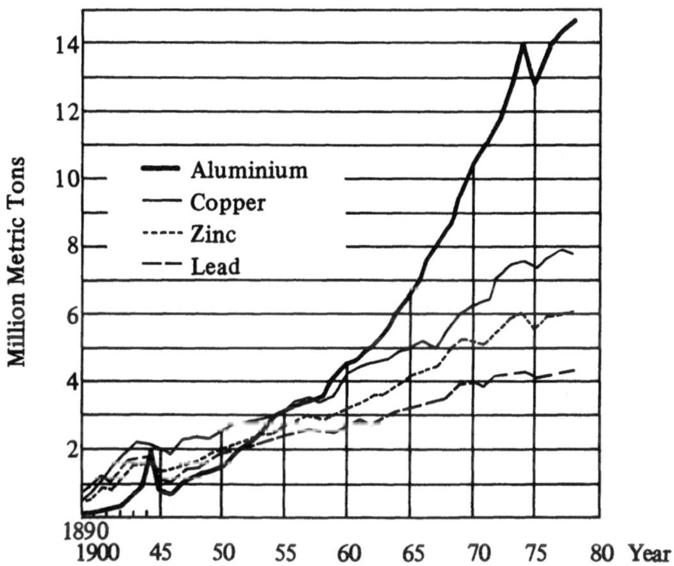

Figure 11: World Primary Aluminium Production from 1890–1978 in Comparison with Other Non-ferrous Metals

Source: Metallgesellschaft AG, Frankfurt a. M.

In 1921, the world's annual production of primary aluminium reached the 200000 ton level; in 1950, 1.5 million tons; and by 1978, about 15 million tons (including estimated production in Eastern Bloc countries). Since World War II, the production and consumption of aluminium has increased at a rate of more than 8% per year, that is, doubling in less than nine years.

Today, aluminium is the second most widely used metal in the world. On a volumetric basis, more aluminium is consumed than all other non-ferrous metals combined. Because of its unique properties, aluminium has substituted much older, established materials such as: wood, copper, iron and steel.

The per capita consumption of aluminium, for several countries, in 1978, is given in Table X.

Table X: Per Capita Aluminium Consumption 1978

Country	kg/capita	Country	kg/capita
Austria	10.7	Portugal	3.1**
Belgium	8.9	Russia	8.6*
Brazil	2.7	Spain	5.6
China (PRC)	0.5*	Switzerland	13.5
France	11.5	Turkey	1.6**
West Germany	19.4	United Kingdom	12.0
Italy	11.0	United States	29.7
Japan	19.6	Venezuela	5.1

* Estimates ** 1977 Values

Total world aluminium consumption (primary and secondary) in 1978 was 19.4 million tons.

Aluminium and the Future

What does the future hold for the aluminium industry?

Past increases of 8% or more per annum in the growth rate of primary aluminium consumption will not continue. The growth rate is expected to be around 4–5% per annum for the balance of this century. There are several reasons for this, like drastic price increases in energy and for equipment to produce primary metals. But, when one considers that today the base is much larger than in the past even modest percentage increases in consumption mean large increases in quantity. Let us not forget that 5% of 15 million tons is equivalent to five new smelters each with a capacity of 150 000 tons per year, each year. Furthermore, additional smelters will be required to replace obsolete capacity.

If we look at forecasts of world consumption of aluminium, from the best available data, we find that a 5% growth rate is rather realistic. What this means by the year 2000 is presented in Table XI.

Table XI: Aluminium Consumption Forecast Million Metric Tons (primary and secondary) at 5% Annual Growth

	1978	1985	1990	2000
Western World	15.3	22	28	45
Communist World	4.1	6	7	12
	19.4	28	35	57

At a 5% growth rate, primary aluminium would contribute approximately 30 million tons to consumption in the Western World and 9 million tons in the Communist World, in the year 2000. The bauxite reserves, aluminium's ore, are sufficient to meet these demands. The balance would be provided by secondary aluminium. Estimates for Communist countries are difficult to make. Their growth rate could be higher than 5%, if their ambitious programs are realized.

3.1 Geographical Distribution

The extraction of aluminium and its fabrication into finished products takes place in a series of successive operations, each largely independent of the other. Generally the various processes are carried out at different plant sites in many countries.

Figure 12 divides the world-wide activities of the aluminium industry, for the mid 1970's, between those which are performed in developing countries and those executed in the industrialized countries. The segment to the left of the dividing line represents the developing countries. Australia is included in the Oceania category. One can see that more than 90% of the bauxite reserves and almost 90% of the bauxite production are in developing countries.

On the other hand, around 1975, we can see that two-thirds of the alumina plants were located in industrialized countries. In addition, almost 90% of the primary aluminium was produced in highly developed countries at this time, where also more than 90% of the aluminium was consumed.

Although the annual increase in primary aluminium consumption (1965–1976) was quite significant on three continents: Africa and Asia, both about 13.5% per annum, and South America 12% per annum, the absolute tonnage increases were relatively small since the starting base was low. Countries in these continents do have a considerable consumption potential. However, for at least the rest of this century the industrialized countries will remain the main consumers of aluminium.

Therefore, when looking at aluminium consumption, it is necessary to take into consideration different market environments. The use of aluminium follows a similar pattern in both industrialized and developing countries, only separated by time. First, there is aluminium usage in household articles, then in the electrical sector as industrialization takes place. This is slightly different between the two groups of countries because in industrialized countries aluminium is replacing copper, whereas in developing countries aluminium is utilized from the beginning, by-passing the copper phase.

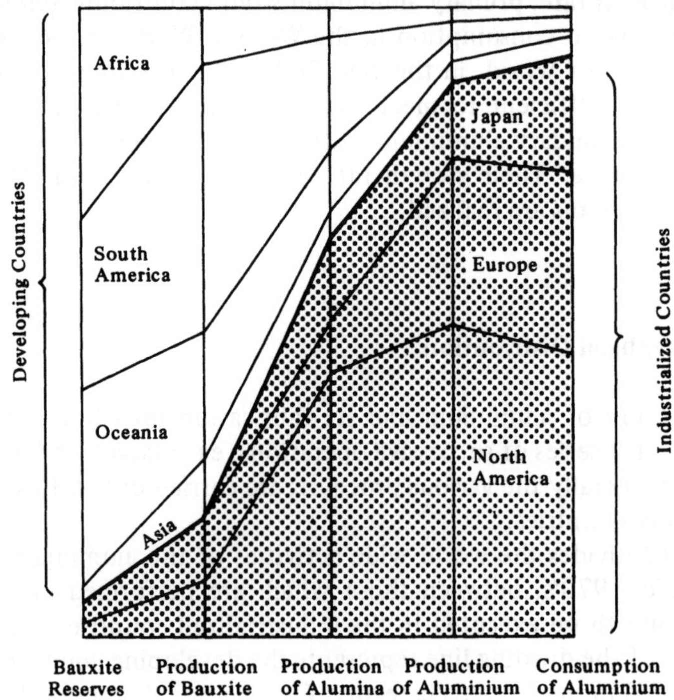

Figure 12: Western World Aluminium Industry Structure
Source: Revue économique de la Banque Nationale de Paris, No. 32, October 1974

Altogether it may be concluded that the latent demand in the developing countries will have an increasing influence on the world-wide aluminium industry as time goes on, affecting both consumption and location of new facilities.

North America and Europe will remain the leading aluminium producers for some time, however, their relative position is declining and will continue to decline. Where aluminium production has become very expensive, inefficient units are likely to be closed down.

It can be predicted that new alumina plants and new aluminium smelters will be located in developing countries on an increased scale, especially in those countries which have bauxite and/or surplus energy. The main reasons for this trend are: forward integration by the bauxite producing countries and the interest of many developing countries to utilize their hydropower potential, as well as the interest of oil exporting countries to make investments in capital intensive growth industries.

3.2 Aluminium's Important Application Sectors

In the industrialized countries of the Western World, aluminium's four major applications are in

- building and construction,
- transportation,
- containers and packaging, and
- electrical engineering,

which currently account for about 75% of total aluminium consumption in the USA and 67% in Europe.

In Eastern countries, the usage of aluminium is quite different. For example, very little aluminium is used in packaging or buildings.

It is obvious that these four application sectors represent the key to the future of the aluminium industry. What happens in these markets will largely influence the entire aluminium industry.

Building and Construction: There are great differences in the construction styles and methods between different countries, and thus in the materials used. Therefore, it is difficult to make generalizations about this market. However, it should be noted that the trend towards assembly line techniques and modular construction has favored aluminium as a building material. Aluminium has made great progress in institutional buildings, such as schools, hospitals, offices, as well as in apartments and real estate developments.

The main competing materials are galvanized and plastic coated steel, PVC, timber and asbestos. Aluminium has some advantage over these materials, because of its good corrosion resistance. It is virtually maintenance-free. In the mobile home market, aluminium is in a stronger position due to its low weight compared to steel.

Other applications for rolled aluminium products are curtain walls, gutters and roof flashing where it competes with copper, galvanized steel and plastics.

In general, the concern for saving energy favors aluminium usage in such applications as insulated windows and doors, solar collectors, heat exchangers, vapor barriers and maintenance-free architectural elements.

Transportation: In most industrialized countries, transportation is the largest application sector for aluminium.

In this sector, the strongest push towards more aluminium use will continue to be found in commercial vehicles, where a kilogram of weight saved is a kilogram of carrying capacity earned.

The application of aluminium castings has been very successful in the past and this trend will continue in the future. The main areas for aluminium castings in transport vehicles are transmission housings, pistons and engine blocks.

Plastic parts are increasing by replacing aluminium in trim applications. Radiators appear to be a very promising application area for aluminium.

Aluminium sheet is used for bodies of commercial or rail vehicles where the power/weight ratio is an overriding consideration.

Competitive materials in transportation are steel sheet, iron castings and plastics.

Bumpers were successfully introduced in aluminium but face competition from glass-fiber-reinforced plastics.

In general, the outlook for aluminium in transportation in specific applications is bright.

Containers and Packaging: Packaging constitutes a substantial outlet for aluminium.

Aluminium thin strip and foil comprise a major part of total consumption in this market. The use of foil is well established for a wide range of packaging purposes. These include tobacco, food and dairy products, pharmaceuticals as well as household applications. Aluminium provides protection against chemical agents, water and vapor as well as against light and ultraviolet rays.

Aluminium can be combined with other materials, for instance plastics and/or paper. Such composites utilize the advantageous properties of each material and offer totally new solutions to packaging problems.

Plastics are challenging the market for aluminium caps and bottle tops. They may also be a potential rival in the tube market.

Beverage cans are a major market for aluminium strip, especially in the USA. There, metal for lids and can bodies comprises nearly two-thirds of total aluminium consumption for containers and packaging. Since the economics of recycling all-aluminium cans are already attractive, aluminium has a bright future in this sector.

Electrical Engineering: In the electrical and telecommunications field, aluminium has made substantial gains.

The present ratio of copper to aluminium regarding metal use for electrical conductors of approximately 3:1 will progressively change in favor of aluminium. If the copper price will stabilize again on a level of one kg copper costing twice as much as one kg aluminium, it can be assumed that the demand for copper will not grow and that the main increase in electrical conductor material will be filled by aluminium.

In the field of high voltage power transmission lines, aluminium in the form of ACSR (aluminium conductor steel reinforced) has substituted for copper. Price advantages combined with reduced weight are the main advantages and no change can be foreseen in the future.

In the field of insulated or covered wire and cable, the current trend to substitute aluminium for copper will continue, especially in power distribution cables.

Application Summary: The future prospects for aluminium in all of its major application sectors are favorable, even at a slower growth rate.

3.3 Mineral Availability

The abundance of present and potential mineral reserves distinguishes aluminium from other non-ferrous metals.

Aluminium comprises approximately 8% of the earth's crust, making it second only to silicon (27.7%), iron is third at about 5%. However, metallic aluminium is not found in nature, it occurs in the form of hydrated oxides or silicates (the latter are for instance kaolin or clay).

The principle ore from which aluminium is extracted is called bauxite after the town of Les Baux in Southern France where the ore was originally discovered. Bauxite is one of the most abundant minerals in the earth's crust.

It is found primarily in equatorial and subtropical regions. The alumina hydrate content of bauxite, expressed as aluminium oxide, (Al_2O_3) usually ranges between 40–60%. In addition to aluminium oxide, bauxites contain varying amounts of water, iron oxide, silicates, quartz expressed as silica (SiO_2) and usually minor amounts of other metallic oxides.

Today's proven bauxite reserves are sufficient for about 300 years at 1978 consumption rates. In addition, there are large amounts of lower grade bauxite containing 30–40% aluminium oxide. These low-grade bauxites can be processed in existing alumina plants. Taken together, it is estimated that there is a sufficient supply of bauxite for more than 1000 years.

In addition to large quantities of lower grade bauxite, there are several other potential sources of alumina besides bauxite, such as: nepheline, alunite, labradorite, andalusite, clay, schist and kaolin. However, most of these other sources have lower alumina contents than bauxite and since 90% of all bauxite and alumina is used in the production of aluminium, bauxite will in all probability continue to be the preferred ore, especially if one considers the large proven bauxite reserves.

As indicated earlier, for bauxite, there is no reason for any supply pessimism ("Limits to Growth Syndrome"). In one decade, 1966–75, about 20 times more bauxite reserves were proven than used and during this same interval the proven reserves increased from about 6 000 million tons to more than 17 000 million tons.

Cartel Action; Since 1974, the International Bauxite Association (IBA) headquartered in Jamaica, a cartel representing countries in which up to now more than two-thirds of the bauxite is produced, has recommended that its members exact a levy tied to the aluminium metal price. This has had the effect of increasing the cost of raw material. However, the resulting price increase has not been detrimental to the competitive position of the aluminium industry.

A drastic increase in the IBA levy on the high-grade bauxite used today is not likely because the aluminium industry would then begin exploiting other raw material sources, beginning with low-grade bauxite and some high-grade clays.

3.4 Primary Aluminium Production

Primary or "virgin" aluminium is produced from bauxite in two stages after the bauxite is mined. First alumina (Al_2O_3) is extracted from the ore, in an alumina plant. Then, the alumina is fed into electrolysis cells where the aluminium is produced, utilizing the Hall-Héroult Electrolysis Process. Figure 13 shows a cross section of a modern electrolysis cell.

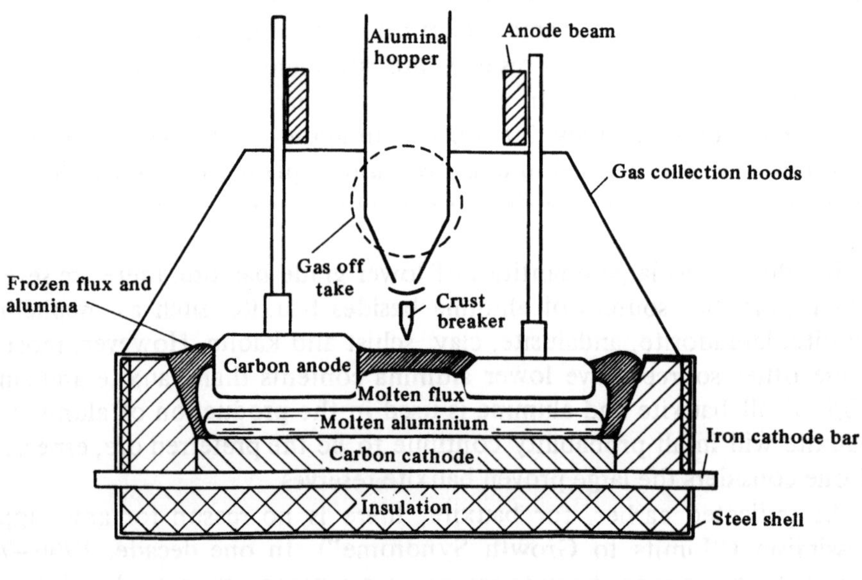

Figure 13: Hooded, Center-Break, Prebake Anode Cell

Source: International Primary Aluminium Institute, London

At a temperature of approximately 950°C, direct current is passed through a current-conducting salt bath in which the alumina is dissolved. The bath consists of fused sodium aluminium fluoride (Na_3AlF_6), commonly called cryolite, or a mixture of cryolite and other fluorides. Because alumina dissolves in the salt bath, the electrolysis takes place considerably below the melting point of alumina.

Under the influence of the electric current, aluminium metal is deposited at the negative pole and therefore, collects at the bottom of the cell from where it is tapped on a periodic basis. Oxygen is released at the anodes where it reacts with carbon, forming CO and CO_2. Thus the anodes are consumed and must be replaced regularly.

Aluminium processing and fabrication are described further in 3.6.

3.4.1 Ecology

Many basic industries, as well as the aluminium industry are faced with ecological constraints or problems today. Because the vast majority of bauxite mines are worked by surface mining, restoration of mined areas is a task which the aluminium industry normally carries out in accordance with regulations in each specific country.

In the production of alumina, bauxite residue ("red mud") is a waste product for which no commercial use has been found despite over 50 years of intensive effort. Therefore, some type of disposal system is normally used. Nevertheless the aluminium industry is constantly evaluating different means of bauxite residue utilization and improved methods of disposal.

Although there are new processes on the horizon, the production of primary aluminium is accomplished by the Hall-Héroult Electrolysis Process today.

The first generation of aluminium reduction plants utilized open cells and, fluorides and particulate material evolved during electrolysis were emitted into the atmosphere. Since then, the aluminium industry has made great strides in reducing emissions so that planners can avoid undesirable ecological impacts by incorporating appropriate pollution control technologies together with proper plant site selection criteria.

New reduction plants now employ hooded cells to direct the evolved material into abatement systems where it is captured and then recycled into the reduction cells.

However, despite the aluminium industry's good record in environmental matters, expansion will be difficult in developed countries due to the heavy concentration of industrial plants already there and a greater awareness of nature's assimilative capacity.

3.4.2 Energy

The production of aluminium is energy intensive. On an industry average, the production of 1 kilogram of aluminium consumes about 15 kWh. But this figure needs to be seen in relation to other users of primary energy. For example, in 1976–77, the aluminium industry used 1.7% of the total installed electrical capacity in the USA. Some additional energy is needed in alumina plants and semis plants, mainly thermal energy. However, the bulk of energy consumption is as electricity for the smelter.

It should also be pointed out that the majority of aluminium production is based on electricity generated by renewable hydropower sources. Furthermore, Table XII shows that there is a vast potential of unused hydropower in Africa and South America.

Table XII: Utilization of Hydropower

	Potential (1000 MW)	Utilization (1000 MW)	%
Africa	437	9	2
South America	288	21	7

As a point of reference, the production of one million tons of aluminium would require an installed capacity of 2000 megawatts or in other words today's aluminium capacity of 15 million tons equals 30 000 megawatts. This means that there is no shortage of renewable hydropower sources which can be utilized for aluminium production.

In addition, the aluminium industry as a whole has done much to reduce energy consumption. Further reduction in the amount of energy consumed is possible. For example the aluminium industry in the USA has voluntarily pledged a 20% improvement in energy efficiency by 1985, with 1972 as a base year. There is much emphasis on reduced energy consumption because energy constitutes a large portion of the cost of producing aluminium.

Although aluminium production consumes energy, the use of aluminium as a material of construction saves energy. Therefore, on a balance, in many applications the use of aluminium results in overall energy savings in comparison to other materials.

3.5 Secondary Aluminium Production – Recycling and Scrap

The secondary aluminium industry, recovery of aluminium from scrap and dross, is assuming greater importance today and will be more important in the future. There are three main reasons for this. First, the amount of aluminium to be reclaimed is greater because of more extensive use in packaging and transportation. Second, because of the long life of many aluminium products (10–30 years) and since aluminium has been an industrial metal for a rather short period of time, material from such applications is just now becoming available for recovery. Third, remelting of aluminium consumes only a fraction of the energy required to produce primary aluminium (this is covered in greater detail in Energy Accounting of Materials, p. 146).

The "ore" for a secondary aluminium smelter consists mainly of old scrap and new scrap.

Old scrap (which was previously a product in the market) constitutes only approximately one-quarter of secondary aluminium production in a given year, the rest is "new" scrap like blanking and trimming scrap and chips or "run-around" scrap from fabricating plants.

The secondary aluminium producer processes scrap and dross with a high aluminium content by analyzing, separating, crushing, drying, melting, refining, alloying and casting. Improved techniques for these process steps have evolved over the years. However, despite the multiplicity of process steps, the investment costs per ton of annual capacity are only a fraction of the investment required per ton of annual capacity for a primary reduction plant. Of course, not all of the methods mentioned are used in each secondary smelter.

The secondary aluminium industry mainly utilizes reverberatory furnaces but also induction and rotary salt bath equipment. Pre-melt cleaning systems help minimize melting losses as well as preserve the alloying constituents contained in the scrap. Traditionally, the majority of secondary aluminium has been used to produce castings. Now there are efforts to upgrade the secondary metal to make wrought alloys as well. When there is a large amount of a specific old scrap product, for example beverage cans in the USA, the metal contained therein can be used again to make the same product. As a matter of fact, can reclamation technology almost comprises a closed-cycle within the secondary aluminium industry in the United States. In 1978, the average life span of an aluminium product, before it would appear as "old" scrap for remelting, was 12 years in Europe and somewhat less in the USA, because of increased recycling of beverage cans with a life cycle of less than one year.

3.6 Description of Aluminium Processing

3.6.1 Alumina Extraction

As stated previously, after the ore is mined aluminium is produced in two stages, the first being alumina production. The term alumina is used to designate anhydrous aluminium oxide (Al_2O_3).

In the Western World, the Bayer process* is by far the most important process used in the production of aluminium oxide from bauxite, even though it has been known since the end of the last century. Since that time the process has been refined and improved. Today, it requires large-scale equipment and heavy investment for economical production. Since each bauxite deposit varies from others in its properties, the methods of treatment differ which means that alumina plants are almost tailor-made for a particular bauxite. Nevertheless, the processes are basically similar and a general description is given in the following.

The bauxite from the mine is crushed and ground. Afterwards it is dried and mixed with a solution of caustic soda in large autoclaves. There, under pressure and at a temperature of 110–270°C the alumina contained in the ore is dissolved to form sodium aluminate. The silica in the bauxite reacts and precipitates from solution as sodium-aluminium-silicate. Iron and titanium oxide, and other impurities are not affected chemically, and being solid, settle out of solution.

The slurry or waste material from autoclaves, known as red mud, is separated from the sodium aluminate solution, washed to recover the caustic soda, and then pumped to disposal areas.

The sodium aluminate from the autoclaves is first filtered and then diluted which at the same time cools the solution to about 100°C. It is then pumped to precipitators. The alumina is precipitated as a hydrate from the super saturated solution through agitation and the addition of seed crystals from a previous precipitation cycle. After vacuum filtering, approximately 50% of the precipitated hydrate is returned to the process as seed crystals. The rest, after washing, is calcined in rotary kilns or fluidized bed furnaces at 1100–1300°C to produce anhydrous alumina.

In Russia, due to a lack of bauxite, a process using nepheline as feedstock has been utilized to produce alumina. Essentially the technique consists of sintering a nepheline ore, or concentrate, with limestone. The resultant sintercake consists of sodium and potassium aluminates and dicalcium silicate. This material is crushed, ground and leached. After leaching, the aluminate liquor is de-siliconized and decomposed by carbonization. Alumina hydrate is separated from the liquor and calcined to obtain alumina.

After evaporation and crystallization, the carbonate liquor yields soda and potash. These are centrifuged, dried and packed for shipment.

Limestone is added to the slime from sinter-leaching to produce Portland cement in a second calcination step.

* named after the Austrian K.S. Bayer

For each 4 to 4.5 tons of nepheline ore (or concentrate) that is processed; one ton of alumina, 0.6–0.8 ton of soda ash, 0.2–0.3 ton of potash and 9–11 tons of cement are produced. This process obviously requires a large market for Portland cement.

It should be mentioned that the Russian aluminium industry now imports large tonnages of bauxite. However, despite the fact that Russia seems to be abandoning their alternate processes there are active programs to develop processes based on abundant raw materials that can be found in almost every country. Acid processes, using hydrochloric and/or sulphuric to digest the ore, appear to be more promising. Included in this group are the hydrochloric acid process developed by Anaconda et al., on the basis of Georgia kaolins and the Péchiney/Alcan process (H^+ process) which uses hydrochloric and sulphuric acids on clays or kaolins, like "Argil", a mineral in abundance in France.

Alunite a potassium-aluminium sulphate, is another possible low-grade ore for the production of alumina. Potassium sulphate and phosphate fertilizers can be produced as by-products. The economic prospects of alunite as an alumina ore will depend on bauxite prices as well as the needs and economics of the fertilizer industry.

However, it is believed that the first alternate materials for today's bauxite will be lower grade bauxites rather than other ores requiring new processes. The main reason for this is economics. Technologies using marginal or low-grade bauxites will require only certain modifications or additional equipment for a Bayer plant and therefore not as high an investment as entirely new plants using totally different processes.

3.6.2 Aluminium Electrolysis

The production of metallic aluminium from alumina takes place in electrolytic cells or pots. Although this process was invented more than 90 years ago, simultaneously in France (Héroult) and in the USA (Hall), it is still the dominant process by which aluminium is produced.

The electrolytic cell consists of a rectangular steel shell lined with refractory material as heat insulation, which in turn is lined with carbon. Carbon blocks in the bottom of the cell serve as the cathode. The cell holds the fused salt electrolyte in which alumina is dissolved. Carbon anodes are suspended from above the cell and dip into the bath. When the cell is in operation the bath is kept molten by the heat generated from the passage of electrical current. The surface is usually crusted over. Alumina is added to the bath as needed by breaking the crust. The aluminium is removed from the bottom of the cell on a periodic basis and is then transformed into useful products.

3.6.3 Aluminium Fabrication

Ingot Casting: Primary aluminium and in-plant scrap are cast into rolling ingot, extrusion billet and wire bar, and to a lesser extent into forging stock in the cast houses of the reduction plants or semis-plants. Appropriate alloying elements are added in the melting or holding furnaces, after which the metal is cleaned and cast. Reduction plant cast houses also produce pigs from a part of the primary metal.

Semis-Plants: Aluminium is provided to the industrial user not only as ingot, but also in many different shapes and semifinished products:

- rolled products, plate, sheet and foil with thicknesses from approximately 150 mm down to 0.005 mm;
- extrusions and forgings in a wide variety of cross sections;
- castings produced by every known method;
- wire and cable.

An aluminium semis-plant may receive ingot or billet for fabrication directly from the reduction plant or from their own remelt shop. The first operation is the hot deformation of the cast ingot at temperatures between 350–550°C. Depending on the process, the deformation may take place by hot rolling, extrusion or forging. Such hot working is often followed by cold deformation such as cold rolling of coils or sheets or wire drawing. Some semifinished products are supplied in the as-fabricated condition in the form of extruded profiles, forged parts and hot rolled sheet or coils. Prior to delivery extruded shapes are usually stretched and straightened which imparts a small amount of cold work to the material.

Mold Casting: In foundries, cast products are usually produced from prealloyed metal supplied by secondary smelters. In some cases, casting alloys are prepared from a primary metal base for products which must meet rigid requirements that can be realized with only minor amounts of impurities.

Mold casting is divided into three groups: sand casting, permanent mold casting, and die casting, which usually produces a finished part in one step. Unlike a semis-plant, a foundry may deliver a finished product which requires no further forming. For this reason, foundries are not usually classified as semis-plants.

3.7 Perspectives for Improved and New Reduction Technologies

Outlook for the Hall-Héroult Process: The traditional Hall-Héroult Process will remain the major process by which aluminium will be produced for many years. The aluminium industry has made improvements in the basic process since its introduction and the effort is continuing. Work is being carried out in the areas of automation, computerization, hooding of cells, improvement in waste gas treatment and recycling of fluorides. Development work is also being carried out on anodes and cathodes for the electrolytic cell.

New Aluminium Electrolysis Processes: One process receiving attention is the electrolysis of aluminium chloride ($AlCl_3$). The process requires anhydrous aluminium chloride which is obtained by chlorinating alumina from the Bayer process. The electrolytic cell utilizes special non-consumable graphite electrodes. The aluminium chloride is dissolved in an electrolyte consisting of a mixture of sodium chloride and calcium or lithium chloride. The process produces aluminium and chlorine, and the chlorine is recycled for chlorination of alumina. The main advantage over the Hall-Héroult Process is an approximately 30% less energy requirement in the reduction step. The cell operates at about 700°C which is a much lower temperature than the average for Hall-Héroult cells (950°C). Alcoa has operated a sizable pilot plant, in the USA, using this process since 1976. However, in 1979, they announced that the process to make aluminium chloride will require another 3–4 years of development before it could be technically and economically proven.

There are numerous other proposed processes which have failed to make the grade on technical or economic grounds. These include direct reduction of bauxite with carbon, a sub-halide process and reduction of aluminium chloride with manganese.

3.8 Industry Outlook

In conclusion, after looking at the facts, we can say that aluminium has a bright future. Why? First of all aluminium is needed in modern society. It is almost irreplaceable in the transportation sector and as a conductor of electricity in transmission and distribution networks.

Second, the aluminium industry is in a fortunate situation. They have abundant raw material and there is no shortage of renewable hydropower sources, for the electrolytic production of the metal.

Third, aluminium can be recycled at a great saving in energy over that required to produce primary aluminium. Aluminium is actually an "energy bank". Scrap can be turned into usable aluminium again with only 5% of the energy required for the production of primary metal. Therefore, recycling of aluminium assumes particular importance as the logical method to economically utilize the energy bank contained in aluminium products.

When we look at the critical attitude which is being displayed towards energy consumption, we can state that this does not constitute a threat but rather an opportunity for aluminium.

4 COPPER*

Copper, utilized by primitive man, is the oldest metal still in wide use. Its history starts some 6000 years B.C., when copper was firstly used from deposits of relatively pure metal. Then, by adding tin, an alloy was produced, after which the "Bronze Age" was named.

Brass, which is a copper-zinc alloy, was introduced approximately 2000 years ago and since then has found many applications.

Copper's main use today is in the electrical field – it has the highest electrical conductivity among common metals. Two other important properties are its high thermal conductivity and resistance to corrosion. The versatility in uses made copper one of the most demanded metals.

Total world production of refined copper, in 1978 was 9.1 million tons (from primary and secondary feedstock).

Out of the 9.1 million tons refined copper, 6.8 million tons (primary, 5.9; secondary, 0.9) were produced in the Western World.

The consumption of refined copper and scrap, in kg per capita, in various regions and countries is given in Table XIII.

Table XIII: Per Capita Consumption of Copper 1974

North America	13.2 kg/capita
Japan	12.0
Western Europe	9.4
Oceania	8.1
USSR, Eastern Europe	4.3
Latin America	1.1
China	0.4
Africa	0.3
Far East	0.14 (excl. Japan)
India	0.13

Source: Intergovernmental Council of Copper Exporting Countries (CIPEC)

What are the perspectives for copper until the year 2000? The world demand for copper (refined metal plus scrap) in the year 2000 is estimated to be approximately 23.5 million tons based on 1977 estimates by the US Bureau of Mines for annual growth rates of around 3.5%. In other words, today's demand is expected to double within two decades.

The above forecast must be treated with a certain reservation. If we consider the evolution of copper production within past years on one

* In cooperation with I. Reznik, Dipl. Ing., Swiss Aluminium Ltd.

hand and today's known economic trends on the other, the forecast of nearly 24 million tons of copper use by the year 2000 appears to be somewhat high.

In Table XIV, a certain slowing down in the growth trend of annual copper production in the Western World during the last 8 years can be recognized.

Table XIV: Copper Production in the Western World (in 1000 tons)

Year	Copper*	Year	Copper*
1950	2 860	1973	6 682
1955	3 323	1974	6 923
1960	4 198	1975	6 272
1965	5 044	1976	6 635
1970	6 163	1977	6 855
1971	5 837	1978	6 780
1972	6 371		

* Refined copper from primary and secondary feedstock

Besides the recession of 1975, there are some other reasons for this slow down, e.g., miniaturization of electrical components and substitution by various competing materials.

Therefore, the US Bureau of Mines projection, which assumes a growth rate of 3.5% in copper consumption for the rest of this century, may be on the optimistic side if steady economic growth does not take place in the Third World.

4.1 Geographical Distribution

Copper deposits are spread world-wide. The important countries, and their mine production as a percentage of the world's total (in 1977: 8.1 million tons) are given in Table XV.

Table XV: Geographical Distribution of Copper Mine Production

USA	17 %	Zambia	8 %
USSR	14 %	Zaire	6 %
Chile	13 %	Peru	4 %
Canada	10 %		

The countries with high copper consumption and their percentage, out of the world's total (in 1977: 9.0 million tons of refined copper) are given in Table XVI.

Table XVI: Main Copper Consuming Countries

Country	% of World Total
USA	22
USSR	14
Japan	13
West Germany	8
United Kingdom	6
France	4
P.R. China	4

4.2 The Main Fields of Application

Copper is a versatile metal, used in many fields. However, due to continuous development of competitive materials, it is subject to substitution in certain applications, e.g., in power lines, by aluminium, and in plumbing by steel or plastic.

The breakdown of applications for the USA is given in Table XVII.

Table XVII: Copper Consumption by Application Sector in the USA (1970–1974)

Sector		Examples
Electrical	54 %	Generators, transformers, motors, wiring
Buildings (non-electrical)	17 %	Plumbing fittings, hardware, roofing
Industrial (non-electrical)	12 %	Heat-exchangers, valves, plates, turbines
Transport	9 %	Heaters, bearings, switches, radiators, wiring
Domestic Appliances	3 %	Washing machines, radios, air conditioners
Miscellaneous	5 %	Chemicals, coins, pigments, cartridge cases

Approximately one-third is used as unalloyed copper metal, and two-thirds as copper alloys such as brass, bronze and others. Copper compounds are used in agriculture as fungicides. On a world basis, this use is probably 1–2% of the total copper consumption.

4.3 Mineral Availability

4.3.1 Reserves and Resources***

According to the US Bureau of Mines (1977) the proven world reserves of copper ores are approximately 450 million tons. In addition, over 1750 million tons might be available from resources that are now subeconomic.

The world's copper resources are shown in Table XVIII in short tons (1 short ton = 2000 lbs = 0.91 metric ton). The first column shows the reserves, based on economically recoverable materials.

The second column shows the resources, not used up to now due to economic and/or technical reasons.

By region, North America and South America each account for approximately 30% of the total world reserves. Africa accounts for approximately 14%.

By country, the largest copper reserves are found in the USA and in Chile: each country has approximately 18% of the world reserves. Other important reserves are found in Canada, Peru, Zaire, Zambia and the USSR, each with approximately 6% of world's total. Actually, these seven countries hold approximately two-thirds of the world copper reserves.

4.3.2 Perspectives

Would these reserves be sufficient for the world copper demand in the coming decades?

According to recent forecasts of the US Bureau of Mines, reserves would be plentiful.

The cumulative demand of copper in the years 1975–2000 might surpass 200 million tons, whereas the already known usable world reserves, in 1977, were 450 million tons, or more than twice as much. Considering today's rate of mining of 8 million tons of copper metal in ores per year and other sources that will be found in the next decades, it has been estimated that copper reserves would last for 50–70 years.

It is difficult to forecast developments beyond the year 2000. Various constraints mainly lower grade ore in combination with higher cost of

* Reserves: Economically recoverable material in identified deposits.
** Resources: Includes subeconomic reserves and other potential sources.

Table XVIII: World Copper Resources
(Million short tons of copper)

		Reserves[1]	Other[2]	Total
North America				
	United States	93	320	413
	Canada	34	120	154
	Other	33	30	63
	Total	160	470	630
South America				
	Chile	93	130	223
	Peru	35	40	75
	Other	22	70	92
	Total	150	240	390
Europe	Total	7	40	47
Africa				
	Zaire	28	30	58
	Zambia	32	70	102
	Other	10	20	30
	Total	70	120	190
Asia	Total	30	70	100
Oceania	Total	20	60	80
Centrally Planned Economies		66	190	256
Sea Nodules[3]			760	760
World Total	Short Tons	503	1950	2453
	Metric Tons	458	1775	2233

[1] Of the listed reserves, approximately one-third of the copper is located in undeveloped deposits. These deposits can move between reserve and resource classifications depending on prevailing legal and economic conditions.

[2] Includes undiscovered (hypothetical and speculative) deposits.

[3] Estimated based on average of 1 percent copper (dry nodules).

Source: US Bureau of Mines, 1977

energy, environmental requirements, host country tax and export policies, would affect output and cost. However, new sources are being explored and new extraction technologies, that may partly offset these increased costs, are continuously being developed. Improved technologies are of special interest for lean ores with copper contents below approximately 0.5%, which are plentiful.

4.4 Extractive Metallurgy

4.4.1 State of the Art

The majority of copper deposits are found in nature as sulphide ores and to a lesser extent as oxide ores, or as metallic copper. The grade of the ore has a great influence on the process to be chosen and the costs of extraction.

In the USA, deposits down to approximately 0.5% copper are presently exploited, mostly by surface mining.

Chile is the only large producer of a rich copper ore (approximately 1.2% copper content) by surface mining. The Chilean ore is therefore a low cost ore.

The African ores are rich in copper (in Zambia and Zaire the copper content is 3 to 5%), but underground mining and inefficient use of labor makes them somewhat less economic.

In recent years, approximately 55% of the world's copper was extracted by open pit mining. In the USA, open pit mining accounts for 90%. Underground mining becomes less and less attractive due to rising mining costs.

The production of primary copper is based mainly on the large scale exploitation of low-grade sulphide mineral deposits.

Sulphide ores are processed into metal in four stages:

1) mining: ore containing approximately 0.4–2% copper is mined;
2) beneficiation: copper-containing minerals are separated from tailings to produce a concentrate containing about 25% copper;
3) smelting: concentrates that contain about 25% copper are smelted to produce "blister copper" of 98% purity;
4) refining: "blister copper" is refined electrolytically to produce "cathode copper" of 99.9% purity. Subsequently, cathode copper is melted and cast into various shapes for fabrication.

Oxide ores are generally leaner in copper content. Due to their different chemical nature they are treated differently, mainly by hydrometallurgical processes. The ores are leached (dissolved) by sulphuric acid. The copper is then recovered from the solution by cementation or by electrodeposition (known as the "electrowinning" process).

Ore enrichment ("beneficiation") and smelting are today already predominantly carried out close to the mine. The refining takes place in the main consuming countries, because secondary input into the refinery is generated mostly by the final users in industrialized countries.

4.4.2 Investment and Production Costs

The main cost component in copper production is mining. Mining involves large investments and high production costs. Exploration by itself is an expensive operation. Some large Anglo-American mining companies estimate that exploration costs amount to 5–10% of their total yearly sales.

The investment costs per metric ton of refined copper are estimated to be in the range of \$ 5000 to \$ 8000 (1977), out of which 74% goes into mines, 17% into smelters, and 9% into refineries. According to the experience of one of the largest US copper companies, because of pollution abatement requirements, the investment for smelters can reach 25–30% of the total investment cost.

The production costs are mainly in mining and preparation (beneficiation) of the ores as shown in Table XIX.

Table XIX: Production Cost Structure

Mining	20 %	45 %
Beneficiation	25 %	
Smelting		25 %
Refining		
Transport (overseas)		
Discovery, Development, General Overhead, Taxes, etc.		30 %

Investments in new copper mines are expensive. Thus, copper companies do not wish to invest in countries with political instabilities and/or trends to nationalize private mining property. This, combined with the previously mentioned environmental constraints for US smelters may soon create a relative shortage and a drastic price rise of copper – which was in fact the case already by the end of 1978.

4.4.3 Outlook

Further advanced exploration and mining technology would assist the copper industry in finding new, usable deposits, and in lowering mining costs.

The mechanization and automation of mining operations in open pit mines as well as in underground mines, would result in lower material handling cost, and increase outputs.

Production technology will be strongly affected by environmental legislation, energy shortage and cost, and ore grade.

Air pollution control has become a major concern of the industry in industrialized countries because of strong legislative requirements. The need to reduce the emission of sulphur dioxide gases, at the copper smelters and refineries, forces industry to invest in new technical controls or alternative processes.

The expenditures involved in pollution controls take a large share of industry revenues. In the United States for example, a study by A.D. Little*, estimates that the total cummulative spending by the copper industry for pollution control (capital and operating cost) in the 10 year period 1978–87 will be approximately 2.0 billion dollars. The total expenditure in 1978 is estimated at nearly $ 160 million (all figures in 1974 dollars).

Various techniques are being developed to capture and control oxides of sulphur (e.g., by new gas collection systems, and by reduction of gas volume). However, whereas these measures are intended for the existing pyrometallurgical processes, the general trend is towards hydrometallurgical processes like "electrowinning" by which copper is recovered directly from leach solutions, thereby avoiding air pollution problems.

Cost or curtailment of energy was not a major constraint for the copper industry until now. But because of increasing environmental control requirements for abatement of air pollution and waste disposal which themselves use energy, energy would become a more important factor in total production costs, especially as leaner ores are used.

Looking towards and beyond the year 2000, it is very likely that the combination of lower grade ores, pollution problems and higher energy costs, would shift the industry more and more into hydrometallurgical processes, and away from present smelting technologies.

Another extraction possibility lies in the exploitation of sea nodules. However, it was not yet been determined whether processes can be established, what the costs are likely to be, and whether the interested private industry can obtain the necessary protection of "mining rights" from national governments or through international agreements.

* A.D. Little: Economic Impacts of Environmental Regulation on the US Copper Industry (1/78)

4.5 Semifinishing

Semifinishing processes include casting into shapes, rolling, forging etc. Most of the processes are similar to those used by the steel or aluminium industry, but adapted to smaller manufacturing lots.

Due to slow growth in copper demand, there is a large overcapacity, resulting in plant closings and retarding of technological development.

4.6 Recycling and Scrap

In the Western World, the total recycled scrap amounts to approximately 38% of total copper consumption, as shown in Table XX.

Table XX: Percentage of Scrap in Total Copper Consumption

Western World	38 %
Western Europe	37 %
Common Market	37 %
West Germany	36 %
USA	43 %
Japan	34 %
Canada	23 %
Other Western Countries	30 %

Scrap comes from various sources:

"New" scrap results from most of the working and fabrication processes. It is used by brass mills for rod production or by copper refineries.

"Old" scrap arises from obsolete machinery and equipment, from old electrical conductors, from the plumbing and fittings of demolished buildings, from cartridge cases etc. It is used mostly by the secondary copper industry.

In the consideration of material supply, only "old" scrap and "new" scrap which is not "run-around" material should be taken into account as an additional source of supply.

Reuse and remanufacturing of copper components has reached large dimensions for small or medium motors and generators from automobiles and household machines. Detailed figures are not available yet. The incentive for the final consumer can easily be realized, looking at a catalogue for spare parts of a European car company: a new generator may cost twice as much as a remanufactured one, using an old but still valid copper winding.

4.7 North-South Dialogue

Copper is one of the important commodities for several developing countries. To some of them, it is the main export item and major source of foreign exchange income.

In Zambia, for example, copper accounts for over 90% of the country's export, in Zaire more than 70%, and in Chile almost 70%. Therefore, the fluctuations in prices and returns cause great concern. This is the main reason why copper was included within the framework of the North-South Dialogue.

In this "dialogue" among industrial and developing countries, the latter are seeking ways to stabilize prices and increase export earnings from their raw materials. One of several mechanisms, proposed by UNCTAD* was to create "buffer stocks" to stabilize prices. Of the 18 commodities considered, copper was chosen as one of the important 10 "core commodities".

However, there are some considerable doubts regarding the chances for success of the buffer stock idea. One of the problems is that the amount of money required for the creation of an effective buffer stock (in the order of one million tons) would be very large.

There are other schemes to assist Third World countries. The Rio Tinto Zinc stabilization scheme envisages the provision of soft loans to Third World countries, in times of low prices, to enable them to sustain their balance of payments situation despite cut-backs in production which would be a condition for the receipt of such a loan. The loans would be repaid during periods of high prices. The scheme seems to avoid most of the pitfalls and practical problems of maintaining stockpiles.

4.8 Summary

Supply and Demand: The world demand for copper in the year 2000 is estimated to be approximately 23.5 million tons.

* UNCTAD = United Nations Conference on Trade and Development

This demand will be adequately covered by today's proven reserves. However, temporary supply and demand imbalances could appear at anytime, accompanied with typical price fluctuations.

Newcomers, New Sources: New copper producers would certainly influence the supply structure. Poland for example is in the course of becoming a large scale supplier.

Countries like Argentina and Colombia, and several Pacific islands, are exploring the possibility to exploit their ore deposits. Eastern bloc countries intend to penetrate the Western copper market within the next years.

In North America, existing substantial resources might be utilized or reopened. Entirely new supply sources requiring new extraction processes are in sight, e.g., sea nodules as mentioned before.

4.8.1 Supply Pattern and Price

The Club of Rome report and others which followed, suggested that resources are scarce and that the future supply of raw materials is in danger. Contrary to these claims, it has been recently established that reserves are more than sufficient and future supply could be secured.

One of the important factors affecting the supply pattern is the trend towards ownership or greater control of mineral deposits by producing country governments. In 1960, only 2% of the copper mining capacity in the Western World was government owned; today, approximately 50%, mostly by developing countries.

The Intergovernmental Council of Copper Exporting Countries (CIPEC) is an example. One of the major objectives of CIPEC, which was established in 1967 by Chile, Peru, Zambia and Zaire, was to stabilize the price of copper so that money for developing new mines and smelters could be provided. At the end of 1974, CIPEC announced their plan to exercise price control by reducing exports. This action failed, because the plan was not unanimously followed by the member countries.

Today, CIPEC controls 50% of the Western World mine production and 62% of Western World's export trade (mine, smelter and refined copper production). However, its effect on stabilization of price has been less significant than expected, as some of its member countries at times act independently according to what they think is in their own best interest.

Supply and demand imbalance – aggrevated by commodity speculations – results in strong price fluctuations. To illustrate these fluctuations the development of market prices is given in Table XXI.

Table XXI: Average Copper Prices*

Year	DM / 100 kg	Year	DM / 100 kg
1968	495	1974	542
1969	566	1975	311
1970	525	1976	361
1971	387	1977	312
1972	349	1978	280
1973	477	1979 (June)	361

* Wirebars, West Germany

Copper is a "price-nervous" metal and its price behavior is unpredictable. It seems therefore that the price and not shortage of reserves is going to dictate the future development of the copper industry.

4.8.2 Present and Future Problems

There are several basic problems affecting the copper industry. The first is supply. The imbalance in the short-term supply and demand relationship is reflected in the instability of copper price. Instead of being able to adjust prices by regulating the production, the world is confronted with a situation where supply is no longer governed by purely economic considerations.

Neither CIPEC nor UNCTAD succeeded in their efforts to solve this problem, and there is no solution yet in sight. In fact, contrary to the argument that price stability might be achieved by international commodity agreement, the lesson of history is that as long as even relatively free markets exist, the operation of market forces is the most efficient mechanism to take care of such situations.

The second problem is future investment and its effect on copper price. In order to produce approximately 20 million tons of copper world-wide in the year 2000, new mines must be opened. The large investment involved on one hand, and the trend of nationalization of mines in developing countries on the other, confronts the industry with difficult questions of not only where would the money come from, but in what parts of the world should it be invested.

Nationalization of mines, and in certain cases lack of political stability, has tended to shift exploration and mine development to politically stable countries. Such shifts can result again in increased mining and operating costs, which is reflected in the copper price.

In the USA, which is the largest copper consumer in the world (approximately 2 million tons), the copper industry is currently weak. The situation is reflected in the report of A.D. Little which describes it as follows:

> "In the past decade, the United States copper industry has experienced modest growth in sales, low return on invested capital, eroding profit margins and higher debt, reflecting the combined pressure of inflation, higher cost of capital, increased capital requirements for environmental control and the worst recession during the postwar period . . . With capital expenditures increasing faster than internal cash generation, the cash-flow position of the companies has deteriorated. Consequently, there has occurred in recent years a sharp increase in external financing. Over the last five years, overall debt has approximately doubled, while equity has increased insignificantly."

4.8.3 Industry Outlook

1) The copper industry world-wide has difficult problems to solve, and a huge task ahead. Surely it takes some bold decisions to disregard the present situation and look forward with optimism into the future.
2) Due to its specific properties, copper will further play an important role, especially in the electrical field.
   Hence, a significant increase of consumption is expected in the next two decades, which might be in the range of 20 million tons by the year 2000 (twice today's output).
3) Currently, the copper industry is beset with problems. Overcapacity in smelting, refining and mainly fabricating, lack of investments for new mines and the cost of air pollution control are only few of them.
4) No wonder, that there are different opinions regarding the future of the copper industry. Some knowledgeable people argue that technically, it is a healthy industry which will overcome the difficulties ahead. For instance, the oil industry is optimistic about copper for the 1980's – it is investing in the copper industry both in the USA and in developing countries (such as Chile). Others claim that the outlook of the copper industry is neither bright nor encouraging, and that the technological solutions offered (such as hydrometallurgy) are not satisfactory.

The conflicting opinions point out the existing uncertainties, and the definite need for in-depth technology assessment and planning of alternative routes, which undoubtedly would permit answering and establishing more clearly the "Quo Vadis" of the copper industry.

5 CEMENT AND CONCRETE

By P. Kelterborn, dipl. Ing. ETH. SIA

Cement is a hydraulic bonding agent used in building construction and civil engineering. It is a fine powder obtained by grinding the clinker of a clay and limestone mixture which was calcined at high temperatures. When water is added to cement, it becomes a slurry that gradually hardens to a stone-like consistency.

Concrete is a construction material consisting of a hardened mixture of cement, water and a suitable aggregate (filler) such as: crushed stone, gravel or sand. Concrete was used in a crude form and on a small scale by ancient Mediterranean cultures. Today, the construction industry consumes an enormous volume of concrete with various components and reinforcements.

Of all the possible cements or binders, Portland cement is the one most widely used in concrete. Natural sands and gravels, with the appropriate grading of the grain diameters, are the dominant fillers. By its very nature, hardened mortar or concrete has a high compressive strength and a comparative weakness under tension. This is the reason why concrete is often combined with a reinforcing element. Steel bars of mostly round cross section and various lengths and thicknesses are mainly used for this purpose. They are generally placed into forms or molds prior to pouring or placing the fresh concrete*.

In its most basic form, concrete today is an optimal combination of a chemically reacting binder, an inert filler and a reinforcing element. At the beginning, the binder slurry gives the fresh concrete a semi-fluid consistency suitable for the pouring or forming process. During the subsequent curing or hardening time, the binder terminates its chemical reaction resulting in a solid product of considerable strength and durability.

For the year 1978, world-wide cement production can be estimated at 780 million metric tons and the total output of concrete at approximately 6900 million tons or 2900 million m^3. This last figure represents the load of a freight train circling the earth more than sixty times at the equator or the volume of a cube of more than 1.4 km by 1.4 km by 1.4 km. In comparison with other materials industries it is worthwhile to note that the world's cement and steel production is, when measured in tons, of the same magnitude.

* The expert reader may excuse that the two words concrete and mortar will stand here for the reinforced as well as for the unreinforced version.

5.1 Components of Concrete

The wide range of materials which can be used in modern concrete and mortar is shown in Table XXII.

Table XXII: Today's Components of Concrete

Binder Systems	Aggregates	Reinforcing Systems
– ordinary Portland cements	– natural sands and gravels	– ordinary steel bars
– blended Portland cements	– crushed aggregates	– wire mesh
– special Portland cements	– natural light aggregates	– special purpose steel bars and steel cables
– above Portland cements combined with concrete admixtures	– pyro processed light-weight aggregates	– asbestos fibers
– diverse other mineral binders, mostly reacting with water	– natural heavy aggregates	– prestressing systems – steel fibers
– polymer binders, mostly two component systems and other special binders	– inorganic and organic wastes – special inorganic and organic products	– glass fibers – plastic fibers – glass fiber tendon systems

Binders: Portland cement is technically and economically advantageous and therefore the most important binder for concrete. As such it has become a key material of our present civilization. It is manufactured in large factories using continuous high temperature processes. The rigid rules of production for high investment, high volume, and low-price materials govern this industry. One metric ton of Portland cement costs approximately fifty US dollars (Jan 78) and the investment for a complete plant amounts to 200–400 million US dollars or 50–100 US dollars per ton of yearly capacity. A modern production line can produce 5000–10 000 tons per day. The average energy input is about 1 000 kcal per kg of cement, a value which depends to a large extent on the process (dry or wet) and the equipment being used.

In broad terms, the production method is simple. It consists mainly of grinding and heating up to a clinkering temperature an appropriate blend of limestone and clay*. After this clinkering phase, gypsum is added and the whole is finely ground into the grey powder known as Portland cement. The universal availability of the mineral raw materials allowed the cement industry to spread around the world, keeping as close as possible to the location of the limestone and clay quarries. Due to the mineralogical characteristics of these raw material sources and due to different concepts in the production processes, Portland cements are not all completely alike but show some variations in their performance and behavior. But nevertheless, today Portland cement can be regarded as a chemically and physically well-defined product and all highly developed industrial countries have established a strict system of quality controls and official performance specifications. Beside these practically unavoidable variations, a considerable range of particularly designed "special Portland cements" serve to answer the needs of the market: high early strength – sulphate resistant – low heat – high alumina – and expansive cements are among the.best known types. In other cases and mostly for economic purposes, pure Portland cement is blended with suitable reactive or inert extenders or fillers like fly ash, blast furnace slag, pozzuolana, etc.

Complete replacement of Portland cement by basically other types of binders (phosphate systems, polymer resins, etc.) are continuously invented and tried out. So far, this has only confirmed that Portland cement cannot be replaced and will remain even without serious competition, due to its excellent cost-performance-availability ratio. One of the interesting developments in this field were the polymer based binders of the epoxy and polyester types. Because of their 15 to 30 times higher cost and their sophisticated application methods, these polymer mortars will mainly be used for concrete repairs, concrete glueing methods as well as for injections, impregnations or protective coatings on cement-concrete and mortar. This is a supporting role and all predictions of serious cement replacement, by these glamorous plastic materials, never had a realistic economic basis. It is worthwhile to remember that plastics are based on petrochemicals and have by nature a much lower temperature resistance than mineral binders.

Aggregates: Natural sands and gravels are the optimal material for most types of concrete. They only need the simple processing of washing,

* Selected clay containing (argillaceous) limestone can be used as well. Besides this commonly used direct method of cement production, the following alternative was once widely used in the USSR: When aluminium oxide (alumina) is produced from nepheline, at the same time several tons of cement could be obtained per ton of alumina (pages 62–63). But because of logistic and economic disadvantages, this indirect method of coproduction was gradually replaced by the more efficient direct method.

screening and sometimes crushing. In a modern large quarry, about 25 million US dollars are invested in equipment and buildings, but an effective operation can already be started on an investment 100 times smaller. As compared to cement, the aggregates much more need to be as close as possible to the final location in order to keep transport costs and logistic problems under control. For this reason, costs on site differ very much, mostly ranging between 5 to 15 US dollars per ton or about 1 cent per kilogram.

The present volume of aggregates needed is about the sevenfold that of cement, or some 5900 million tons a year. Lightweight aggregates (natural or pyroprocessed minerals) are today a most frequently used speciality. Reduction of weight and improved thermal insulation are the advantages justifying the higher costs. For factory-made cement products, organic materials like wood chips, or polymer foam pieces, can serve similar purposes. Up to a certain proportion, many kinds of ground or crushed industrial wastes are also mixed into concrete and it is not always easy to decide whether the function as aggregate or as practical waste disposal is more dominant and economically advantageous.

Reinforcing systems: Steel reinforcements in all well known forms (round steel bars, wire meshes, prestressing units, etc.) provide more than 99% of the reinforcing systems of concrete. The reasons are evident.

– The support of a well established steel industry with a mature technology.
– The Portland cement concrete, being an alkaline material, provides excellent rust protection and a strong mechanical adherence to the steel.
– Steel and concrete have the same coefficient of thermal expansion, thus allowing the great durability of the composite material concrete.
– There is no serious competing system in sight with a similar cost-performance-availability ratio.

One remarkable reinforcing speciality needs to be mentioned here because of its maturity, excellent performance and global distribution: asbestos fibers. Extremely thin asbestos fibers are mixed into a fresh cement slurry which is then further processed and finally formed in simple molds or with sophisticated mechanical equipment. Then the product is specially treated and cured. Corrugated plates and pipes, amongst others, have gained universal acceptance in the building industry.

5.2 Energy and Environmental Aspects

Of interest are the subjects: energy consumption, air pollution, workers protection and finally the esthetical-biological conservation of nature in the areas of raw material mining for limestone, clay and aggregates.

By itself, the energy need of the cement industry is considerable. A large amount of heat for the clinkering furnaces must be provided by fossil fuel of that type which is most economically available in the local region. In highly developed countries the share of the total national energy consumed for cement manufacturing is however only in the order of a few percent. Recently, waste heat from cement plants is more and more utilized to produce steam, electricity or to heat suitable buildings directly.

With modern air filtering techniques and various other improvements in the grinding and materials handling methods, a cement factory is able to keep even its "short term dustfall quota" on the immediate surroundings at a tolerable level and well within the unavoidable nuisance values of other accepted heavy industries.

General workers protection in the cement and concrete industries is now an accepted standard procedure. It is important to completely prevent ear and lung damage. Fully satisfactory results are obtained by well designed dust and noise controls inside the raw material quarries, the manufacturing plants and on the construction sites.

It has become commonly recognized that depleted mining areas are not pleasant to look at. Large area recultivation of these zones with grass or trees poses no serious technical or economic problems.

In the perspective of the many known major energy and environmental problems of our time, it must be said that the cement and concrete industry, although not completely free of difficulties, is not a significant part of them. It is generally accepted that energy and environmental improvements require a considerable financial investment and an economy which can afford this effort.

5.3 Today's Cement-concrete

The manufacturing process of a concrete structure or a concrete product goes through the following typical steps.

- Supply of all the raw materials to the mixing plant. In view of the total annual volumes and weights involved, this by itself is a major nationwide transport operation. It is however so much decentralized that it completely escapes attention.
- Proportioning or dosing and thorough mixing of cement, water, chemical admixtures and aggregates. Because now the fresh concrete must be

used within hours, the mixing machine is mostly located directly on the job site, in the prefabrication plant or in the ready-for-use concrete plant. Concrete can be mixed by hand. Modest mechanical installations cost only a few 1000 US dollars. A well equipped mixing plant for a daily production of up to 500 m^3 needs an investment of approximately 0.5 million US dollars while a large ready-mix concrete plant, without transport vehicles, costs at least 5 million US dollars.

- The formwork or molding and reinforcing work. In general, timber or metal is used to make the negative mold or form into which the fresh concrete can be placed and usually, the steel reinforcement is fixed into the formwork before casting. Specially in prefabrication, these forms can be re-used many times resulting in an appreciable economy.
- The casting and compacting operation, bringing the concrete into the molds or into its final shape and location. In this phase, many countries have introduced standard quality control procedures, thereby using internationally accepted specifications.
- And finally the curing or hardening process. If left to the normal weather, concrete hardens perfectly, but curing (steam curing, vacuum curing, etc.) has developed into a separate technology, greatly influencing not only the hardening time but also many final characteristics of the product.

Because of the rich choice of materials as described in Table XXII and also due to many possibilities to influence the quality of the final result during the five steps of concrete manufacturing mentioned above, concrete has become one of the most flexible, versatile and adaptable materials in the hands of the contractor or designer. It is as well suitable for many low technology conditions as for sophisticated high technology projects. Nearly all operations can be executed by hand, but if desirable, the whole sequence can be mechanized and automated. In contrast to the high investment – highly centralized cement production and the medium centralized aggregate supply, concrete has a unique place in the materials industry: in total volume by far the biggest of them all, in production it is completely scattered into a widely dispersed, relatively small scale and little visible operation.

The basic facts and figures about concrete are given in Table XXIII. The experienced reader will note that average values are presented.

Table XXIII: Concrete Today: Facts and Figures* 1978

Economic Aspects of Concrete:				
Total world concrete production	6900 million tons or 2900 million m^3			
Total world cement production	780 million tons			
Per capita world production (Total population at 4.3 billion)	1600 kg/pers. or 0.7 m^3/pers.			
Per capita production in:				
highly developed countries	4000 kg/pers. or 1.7 m^3/pers.			
developing countries	700 kg/pers. or 0.3 m^3/pers.			
General distribution	Europe	36 %	Latin America	8 %
	USA and Canada	10 %	Asia	25 %
			Africa	3 %
	Russia	17 %	Australia	1 %
In highly developed countries:	25 % of concrete is precast, 75 % is cast in situ. 40 % is ready-mix, 60 % is mixed on the job. Of precast and ready-mix concrete, one-third uses chemical admixtures regularly.			

Cost Structure of Concrete: in US dollars as of 1 January 1978

Steel reinforced concrete, cast and cured in place, for an average industrial construction	\$ 100.–/m^3		
Cement, ex factory, in bulk	\$ 0.05/kg		
Concrete cost analysis:		Cement cost analysis:	
binder	15 %	wages	25 %
aggregates	25 %	energy	45 %
reinforcements	30 %	raw materials, plant, etc.	30 %
formwork, placing and curing	30 %		

Technical Data of Concrete:

Cement content	200 to 350 kg/m^3
Weight (after curing)	2.4 t/m^3
Setting time	2 to 4 hours
Final compressive strength	400 kg/cm^2 (75 % of which is reached after 28 days)
Final tensile strength	20 kg/cm^2 (75 % of which is reached after 28 days)
Durability, service life	10 to 50 years
Abrasion resistance	same range as natural stones
Chemical resistance	excellent against sea water, alkalines, oils and fuels (no long-term resistance against acids)
Fire resistance	excellent

* All data is approximate and must be used for general or comparative purposes only.

5.4 Possibilities and Restrictions of Growth

5.4.1 Economic and Technical Developments

Experience shows that cement consumption per capita increases with a higher GNP. Based on this, the economic future of cement and concrete is indeed promising. Already in many Third World countries in the take-off stage, the cement consumption has started growing rapidly today. In these areas, the decision needs careful study, whether to import the cement, only to import clinker for local grinding or to produce locally. Deep sea port capacity, capital availability and the regional raw material supply are among the most important factors.

Even for a mature country like the USA, some predictions mention a growth rate of 3–4% per year during the next two decades. Such a growth rate would mean that between now and the turn of the century more cement will be consumed in this country than in its whole past!

All the many future technological improvements of the components: binders, aggregates and reinforcements (which are too diversified to mention) will directly improve the characteristics of concrete. Other major impulses will come from expected progress in the field of equipment and working methods. The conclusion is therefore justified that the potential for technical improvements of concrete is as high as it is economical.

It cannot be denied however, that this positive economic and technical outlook is based on the assumption of a peaceful, stable and well-balanced development of the world economy.

5.4.2 Practical Restrictions

These are some of the more important limiting factors to be taken into consideration.

- Energy: The rising cost of energy is considerably changing our whole general picture of the future: The hope of an energy- and money-rich affluent society, surrounded by highly sophisticated technologies will not materialize. The building activity, and in particular cement production, cannot escape from this over all emphasis on energy conservation. This includes also transport and materials handling, the hidden energy user in the construction industry.
- Resources: Of all the materials involved, natural sand and gravel will first start to be scarce and rise in cost, because suitable natural deposits are limited. There are also some countries which lack the basic three ingredients (limestone, clay and gypsum) to produce Portland cement. This creates the necessity for large scale imports, like today in many west African countries.

- Laws and Standards: The national differences or sometimes contradictions in design standards, performance specifications, government codes and legal safety and warranty requirements are a considerable obstacle for efficient implementation of progress into the market. For example, a further growth of prefabrication depends not so much on new techniques but more on uniform integrated design standards, better planning and decision procedures and a much wider public acceptance of standardized shapes and modulized sizes.
- Human Factor: People's priorities or judgements will always be at the root of economic, financial, political or social developments. And, because the construction industry is only one of the means to react to these priorities, the future development of concrete is strongly coupled to this unforeseeable element of human priorities.

5.5 Outlook

The questions are: "What will survive?" and "What will disappear?" or even "What will come new?"

What follows here must be regarded as educated guesses about probable trends.

5.5.1 Construction Methods and Equipment

The main emphasis is on time saving and more efficient use of labor. Therefore, ready-mix concrete, prefabrication and concrete pumping will continue to spread. Or very cheap, simple, hand operated production tools will be introduced in the developing countries. Also dry premixes for many types of mortars will not only be more and more available in bags, but also in bulk (silos and containers), requiring specialized transportation and handling equipment.

5.5.2 Binders

Ordinary Portland cement will remain the basic material and its quality and world-wide availability is the key factor of concrete technology. In view of the 780 million tons produced annually it is hard to imagine how other binders could reach a comparable importance. Future progress is to be expected in the following areas of Portland cement production.

- Energy saving and capacity increasing improvements in manufacturing methods*.

* An interesting example is the partial pre-heating of the raw materials before the actual clinkering phase (new precalciner kiln, developed in Japan)

- Reduction of the production tolerances and other variables for each accepted standardized type of Portland cement.
- Blending with industrial or even agricultural wastes, provided that the speed of hardening and the good corrosion protection for the reinforcing steel of ordinary cement concrete can be maintained.*
- But what designers and contractors need most is an increased speed of the hardening process, reduced shrinkage and an improved density and watertightness (to resist the increasing corrosive atmosphere of the industry contaminated areas and the growing attack of the de-icing salts and frost-thaw cycles). As an answer to these many needs, and as a logical compensation of the trend to more uniform but fewer types of cement per production plant, chemical admixtures to concrete and mortar will remain the most efficient solution. In small dosages they can give back to each strictly standardized cement the necessary flexibility to adapt itself to the varying needs of the job site. The key factors of concrete, which can be individualized, are:
  - acceleration or retardation of setting and hardening time;
  - plastification of the fresh concrete;
  - waterproofing;
  - shrinkage compensation;
  - adhesion power of fresh concrete or mortar to old cementitious surfaces.

For low technology as well as for high technology countries it is therefore obvious that by far the biggest future innovation and diversification potential for the concrete binder is in the combination of cement and chemical admixtures. An important indication, already gaining momentum, is for instance the general acceptance of the new superplasticizers. They result in hitherto unknown strength gains and hardening time reductions or in extreme increases of workability ("liquid concrete"). Present superplasticizers are synthetic, organic products of a definite chemical composition (sulfonated formaldehyde condensates).

The final effect of the above trends will be a reduced number but an increased size of future cement plants and (or combined with) a reduction of the different types and grades of cement which each is producing.**

* In many coal-burning, highly developed countries, fly ash is available in a comparable volume to the amount of manufactured Portland cement. This fact alone already illustrates the economic importance of the issue. In some developing countries agricultural wastes can be used: rice husk, for instance, can become a promising additive of hardening properties (pozzuolanic characteristics). It can also be processed as a raw material for cement due to its high silica content.

** For some less developed and very large countries (like the People's Republic of China) the location of small factories widely distributed over the area might be optimal until a certain regional volume of consumption has been established. After that growing period of mini-cement plants, the emphasis goes back definitely to larger units.

5.5.3 Aggregates

A tendency already visible is backwards integration of the cement and ready-mix concrete industry: To secure their market, they both have started to purchase on a large-scale their own gravel quarries.

Because natural aggregates will gradually become a bottleneck and industrial waste disposal a growing public concern, the coupling of these two can be expected. A new kind of waste disposal technology will develop.

5.5.4 Reinforcements

Recent research in fiber reinforcements (minerals, glass, plastics, steel) received considerable publicity and created a level of exaggerated expectations. Costs, sophisticated application techniques for placing, molding and curing, and the brittleness of the final concrete or mortar products indicate that this fiber technology will not replace steel reinforcements for general use. In specific and small market niches it will be interesting to watch to what extent the substitution of steel by fibers will continue. Specially the alkaline-resistant glass fibers, developed mainly in the UK and the USA, are of importance. The main application is the production of thin walled, lightweight beams, plates and other structural or architectural elements. With its low weight, good corrosion properties and high fire resistance, glass-fiber-reinforced cement products will compete with steel elements and fiber reinforced polymer resin products. For special field conditions, fiber reinforced gunite and glass-fiber-reinforced ready-mix mortars will certainly be applied on a growing scale. Apart from the above, the considerable efforts being made to replace asbestos (health risks under investigation) give another impulse to the further growth of fiber technology.

5.5.5 Conclusion

The main emphasis will be on improving the present state of the art by saving labor, energy and natural raw material resources. A large number of specific innovations for selected specialities will continuously be brought forward, maybe not influencing the world economy too much on the whole, but testifying to the creativity of the people involved in the cement and concrete industry.

6 PLASTICS*

Plastics are high polymer, organic compounds. In the beginning they consisted mostly of natural materials like cellulose, proteins, rubbers and resins or derivatives thereof.

Today the main plastics are derived from hydrocarbons which are polymerized with or without chemical modification, e.g., with chlorine, fluorine, nitrogen, oxygen, silicon, sulfur. These are generally referred to as "synthetics". Using various processes, plastics can be fabricated into a number of shapes and forms, for example: profiles, panels, sheets, films as well as molded products.

At first, the synthetic plastics were considered as a subsitute for natural materials, such as wood, natural resins or fibers. Now, plastics are considered a separate group of materials with their own unique properties. There are several hundred individual polymers today, with widely varying properties. However, all plastic materials fall broadly into three distinct categories: thermoplastics, thermosetting materials and elastomers. Thermoplastics are materials which can be softened repeatedly by the application of heat whereas thermosetting materials undergo a chemical change when heated and shaped. As a result they cannot be reshaped thereafter by the application of heat. Elastomers are rubberlike materials most of which undergo a thermosetting reaction, during processing, called vulcanization.

Petroleum fractions and natural gas have displaced coal as the primary feed stock for the plastics industry. Naphtha, obtained from crude oil distillation, is cracked to provide useful intermediates, such as propylene, ethylene and acetylene. The ethylene is used in the manufacture of polyethylene or, in another process, to produce vinylchloride, the basis of polyvinylchloride. Aromatic hydrocarbons such as benzene, toluene and xylene are intermediates used in the manufacture of important plastic materials like phenolics and polystyrene and auxiliary products such as plasticizers and stabilizers. The aromatic hydrocarbons are also the starting point for many polymers utilized in the synthetic fibers industry.

Today, plastics have a significant role in industry and their importance is expected to grow in the future. Perhaps the next century will be called the "plastic's age" because plastics will displace materials used today as steel and aluminium displaced wood many years ago.

* In cooperation with Dr. G. Friese, Swiss Aluminium Ltd.

6.1 General Survey

Within the scope of this book we will concentrate on the plastics category and not go into synthetic rubbers and elastomers or synthetic fibers. Within this group of plastics or synthetic polymers we have chosen to discuss some of the so-called bulk thermoplastics, such as polyvinylchloride, high and low density polyethylene, polypropylene and polystyrene. Figure 14 shows the short, medium and long-term application for these bulk plastics in West Germany, in 1973.

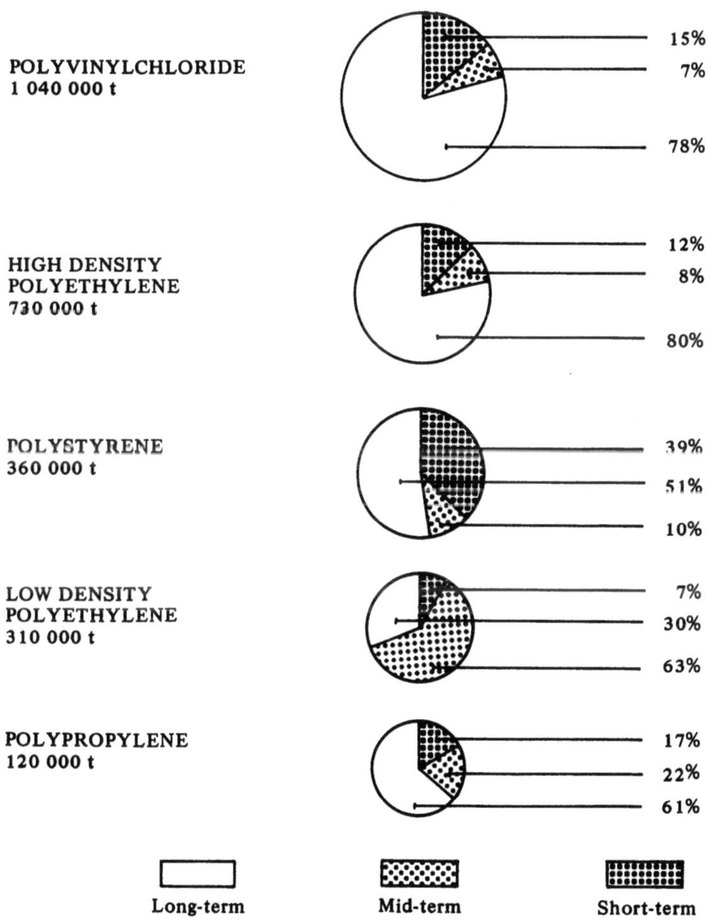

Figure 14: Long-term, Mid-term and Short-term Applications of Bulk Plastics in West Germany, in 1973

Source: Schriftenreihe des Fonds der Chemischen Industrie, Heft 14, 1978

Short-term application is identical with the use of plastics for packaging; medium-term for products lasting several years like toys or household goods; and, long-term for products such as tubing, cable insulation or building components like window frames.

New uses and applications are being found for plastics almost every day and as a result they are part of the new generation of materials displacing metals, natural materials, like wood, as well as classical building materials, like cement and bricks, in certain applications. Examples of these uses in a few sectors are given below.

Building Construction: doors, window frames, wall coverings, façades and curtain walls, pipes, as well as load bearing members.

Machinery and Equipment: housings, enclosures, cases, covers, handles and insulation.

Electrotechnics: insulation material, jacketing, housing, cables, switches and plugs.

Engineering Construction: tanks, reactors, towers, pipes, conduits and valves.

Agriculture: pipes, tanks and green house films.

Automotive: coverings, seats, upholstery, instrument panels and body parts.

Today, the plastics industry is an important part of the world's economy. The production of plastics (excluding elastomers and synthetic fibers) was 41.5 million tons in 1974. This was about 3 times more than aluminium consumption in the same year. However, compared with steel consumption of 707.6 million tons in 1974, consumption of plastics seems relatively small (Table XXIV)

Table XXIV: World Consumption of Selected Materials from 1955 to 1975 (in million tons)

	1955	1960	1965	1970	1974	1975	Average Growth: % per annum
Plastics	3.3	6.9	14.5	30.2	41.5	35.2	+13.0%
Synthetic Rubber	1.0	1.9	3.7	5.6	7.3	6.6	+ 9.8%
Synthetic Fibers	0.3	0.7	2.0	5.0	8.1	7.2	+17.2%
Steel	269.2	345.0	455.0	426.8	707.6	642.7	+ 4.4%
Aluminium	2.7	3.8	6.5	10.2	13.9	12.7	+ 8.1%
Zinc	2.4	2.5	4.0	4.9	5.6	5.1	+ 3.8%

Keeping in mind that the density of plastic is much less than the density of steel, then on the basis of volume and value, plastics consumption is on the same order of magnitude as steel consumption. In the past, the growth rate for plastics averaged about 13% per annum whereas that for steel was only around 4.4%. For the future it is estimated that the growth rate of plastics will be greater than 4% whereas that for steel will drop to around 2%.

To be successful, a plastic component must be designed according to the specific properties and not only be a copy of the shape used with the former material. Figure 15 shows a further customary classification of plastics according to their properties, that is, whether they are thermoplastic, thermosetting or elastomers.

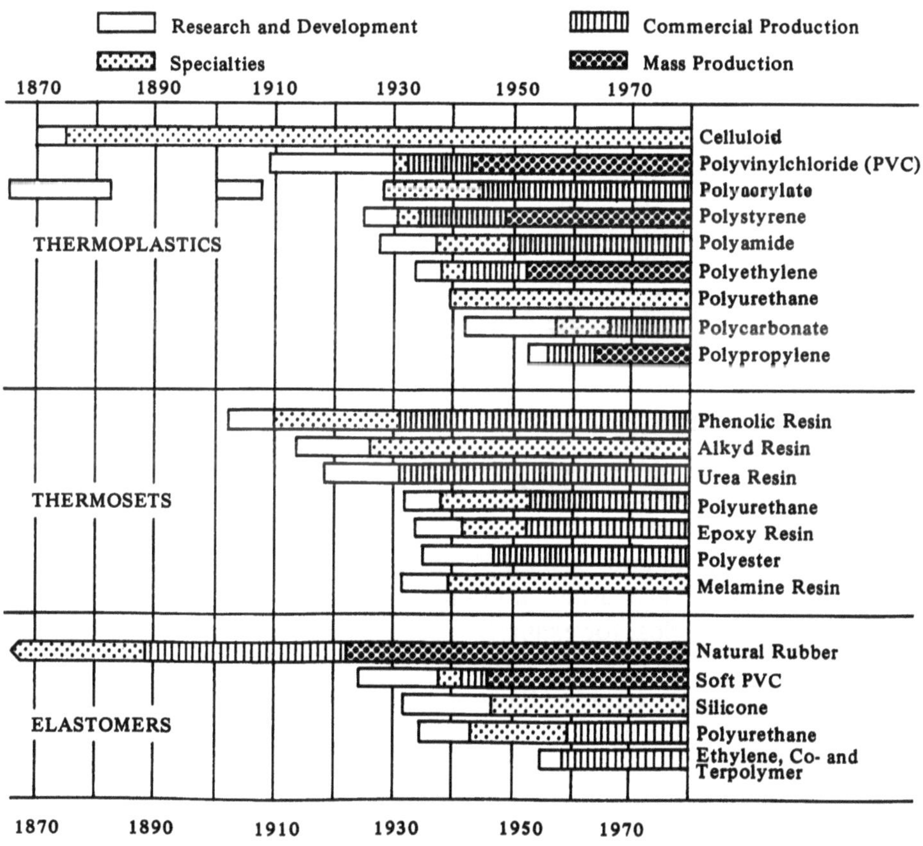

Figure 15: Timetable for Research, Development and Production of Selected Polymers
Source: Schriftenreihe des Fonds der Chemischen Industrie, Heft 14, 1978

6.2 Consumption and Perspectives: Overview

The world consumption of plastics, from 1960–1976, is given in Table XXV. There it can be seen that 5 bulk products: polyvinylchloride,

Table XXV: World Consumption of Plastics from 1960 to 1976 (in million tons)

	1960	1965	1970	1974	1975	1976
Low Density Polyethylene	0.8	2.3	5.5	8.0	7.2	8.6
High Density Polyethylene	0.2	0.7	1.8	3.3	2.9	3.6
Polypropylene	0.02	0.3	1.2	2.5	2.4	3.4
Polyvinylchloride	1.4	3.3	6.1	9.1	8.0	9.2
Polystyrene	0.6	1.2	2.2	3.3	3.0	3.5
Other Polymers	3.9	6.7	13.4	15.3	11.7	15.2
Total	6.9	14.5	30.2	41.5	35.2	43.5

polypropylene, polystyrene and high and low density polyethylene, account for the lion's share of consumption. It can also be seen that all other synthetic polymers accounted for only about 30% of the production in 1976 or 15.2 million tons. In the following sections we will essentially concentrate on these 5 major categories of bulk plastics.

6.3 Geographical Distribution

For all practical purposes, plastics are produced and consumed in the industrialized countries. These, in the order of their importance, are: Japan, USA, Europe, and Eastern Bloc countries. Recently, plastics are being produced in increasing amounts by industrializing countries in Asia and South America, for example, Taiwan, S. Korea, Brazil, Argentina, Venezuela as well as in a few developing and oil producing countries.

The production capacity in Eastern Bloc countries is being expanded rapidly. Also, PR China plans to increase their plastics production to a great extent.

The production and economics of bulk plastics undergoes cyclical changes. At the present time overcapacity prevails. New capacities, principally in Eastern countries as well as in developing and industrializing countries, put additional pressure on the market. There will probably be no change in this situation over the short-term.

6.4 Main Application Sectors

There is practically no sector of man's life without synthetic polymeric materials.

The most important sectors are:

- transportation
- building and construction
- furniture
- household appliances
- sports
- leisure time
- clothing
- packaging
- advertising
- medicine and surgery
- communications
- aerospace
- equipment and plant construction
- electrotechnics and materials handling

In each individual sector many different high polymer materials are utilized and these again in a great variety of types. Research and development continuely opens new application sectors.

The broad application possibilities for polymeric materials are based on the variety of their properties. Almost any combination of properties can be produced through goal-directed development work. The variety is based

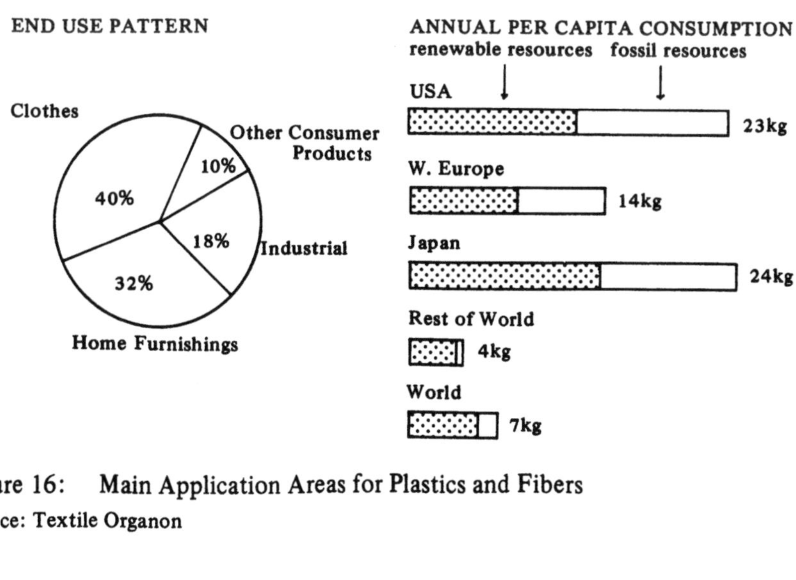

Figure 16: Main Application Areas for Plastics and Fibers
Source: Textile Organon

on a combination of physical and mechanical properties and simple working and forming possibilities. Many combinations of properties are obtained through modification of the molecular structure, changing the morphology, mixing different polymeric materials and combinations with other materials.

Figure 16 shows the main application areas for plastics and fibers as well as the per capita consumption in the USA, Western Europe, Japan etc. and divided according to renewable resources and fossil resources.
The automobile industry is an outstanding example of how varied the application of plastics is. Therefore, it can be assumed that there will be further increases in consumption and production of plastic materials. In the USA for example the following average annual growth rates have been estimated for 1978–1983: polystyrene 6%, low density polyethylene 7%, polyvinylchloride 7.5%, polypropylene 8.5% and high density polyethylene 9.5%.

On the basis of physical volume, plastics production is expected to surpass the production of iron and steel by the mid eighties and in the year 2000 it will be greater than all metals combined.

6.5 Production of Bulk Plastics

Although we have limited our discussion to bulk plastics in the previous sections of this chapter, we still have materials of entirely different characters which require varied starting materials which are produced by diverse processes.

6.5.1 Raw Materials

Today, the only really important feed stocks for the plastics industry are petroleum fractions and gas. The industry is based on the following intermediates: ethylene, acetylene, propylene and benzene as well as chlorine and hydrochloric acid. By far the most important of these is ethylene which is normally produced from naphtha in a cracker. In this process acetylene and propylene are the usual by-products. The starting materials can also be produced from natural gas, which in contrast to Europe, is very frequently the case in the USA. Benzene is produced from crude oil fractions but can also be obtained from coal tar. Chlorine and hydrochloric acid can be produced electrolytically from a sodium chloride solution.

The chemical industry utilizes about 6% of the petrochemical raw materials and 1/3 of that is utilized for plastics. Even in the event of an oil shortage, the chemical industry should be able to obtain this small amount of material because they can afford to pay more than most competing sectors, because of the higher added value of their products. Generally in Europe and a few other parts of the world, with the exception of the USA, utilization of natural gas as a raw material has just begun. This gives another raw material source for the future. Also, the industry can revert to coal as a raw material source. The necessary technology is available since, previously in Europe and today in South Africa, some raw materials for plastics are obtained from coal. All it requires is a modernization and improvement of well-known processes. In addition, in the future, biomass could be utilized as a starting material. In Brazil, for example, due to a shortage of fossil fuels, biomass could play an important role for the chemical industry in the future. However, when looking at feed stocks for the plastics industry in the highly developed countries, for the next one to two decades, the only alternative to oil and gas, is gas and oil.

6.5.2 Manufacturing

Depending on the plastic and the intended processing technique, plastics can be supplied as granules, powders or liquid raw materials.

The thermoplastics generally consist of powder or granules (pellets) in polymerized condition, which can then be formed under heat and pressure.

The intermediate products for thermoset plastics are either granules or liquid substances. In contrast to the thermoplastics, a chemical change takes place under heat and pressure of the forming operation used for thermosetting plastics. Consequently, cured thermoset plastics cannot be reshaped by the application of heat.

6.5.3 Preparation and Forming of Plastics

Only a few plastics can be used in the form in which they are produced, directly for the production of a finished product. Physical properties, workability and application essentially are determined through additives like stabilizers, lubricants, fillers, pigments, plasticizers etc. Before processing, the resins appear as suspensions or emulsions, as well as chips or powder. The polymers in this condition are then compounded through intimate mixing with the necessary additives.

For all conversion processes the polymeric materials are brought to a liquid, semisolid or rubbery stage by heat. The mechanically given shape in a mold or die is retained for thermoplastics only by cooling. Thermosets, which undergo a chemical reaction, may be taken out of the mold hot.

In both cases dimensional allowances must be made for shrinkage due to cooling and/or chemical reaction.

Commercial processes include: compression molding and laminating for thermosets; injection molding, extrusion, calendering, blow molding, and film forming for thermoplastics.

Foamed plastics can be produced through the help of a chemical blowing agent as well as by other methods. Of special interest are the thermoplastic polystyrene foams which are utilized for thermal insulation at ambient and low temperatures and thermosetting polyurethanes used for thermal insulation and cushioning materials.

To expel included air – for thermosets to avoid formation of bubbles due to chemical reaction; and to move the very viscous plastic mass – presses are of great importance in all processing operations.

Thermoplastics are melted in a cylinder and mixed and pushed forwards with a screw, into a closed mold for injection and blow molding, or through a die for profile, sheet and film extrusion. Special cooling devices like cooling dies, cooling rolls, water baths and air currents are needed to maintain the shape of the extruded product and cool it down fast. For packaging films or electrical insulating materials two or more plastics may be extruded together to form sandwich films. For special applications other materials may be coated with the plastic melt coming out of a film die: fabric coated with plasticized polyvinylchloride or polyurethane to obtain leatherlike materials, polyethylene on aluminium foil for packaging film and for cable wrapping.

In compression molding of thermosetting plastics the molding powder or preformed pellet is thermally softened in the mold itself or in a special cavity from where the molten plastic is pressed into the mold cavity, where it hardens (transfer molding).

6.5.4 Post-forming Techniques

Molded articles or semifinished plastic material may be shaped by techniques which are also used for other materials, for example, bending or folding, cold or hot forming and drawing, stamping, sawing, milling, engraving and polishing. For joining plastics, the techniques are: various welding systems, adhesive bonding, screwing, riveting and nailing.

6.6 Special Plastics

In addition to the bulk thermoplastics previously discussed we should mention a few additional members of this group: polyamides (nylons), polymethacrylresins (Plexiglass® *), polyvinylidenchloride, polyacetals (Delrin®, Ultraform®) as well as the heat resistant fluorinated resins (Teflon®).

* ® = Registered Trade Mark

In the thermosets, the melamine resins which permit production of light-colored moldings and electrical insulating parts with high arc resistance should be mentioned. The epoxies used for structural adhesives, chemical resistant coatings and as casting and laminating resins are well-known products. The elastomeric silicones, which contain silicon atoms in their molecular structure, are easily processable and withstand continuous service temperatures of about 200°C without the disadvantage of cold flow under pressure like the fluorinated plastics. Polyester resins will be cited in combination with glass fibers. Even though natural rubber is still an important non-synthetic elastomer, the artificial rubbers have gained the main part of the rubber market. Besides the oldest butadine (BUNA®) polymers which have been modified with styrene (SB Rubber), newer polymers do exist, based on acrylonitrile (nitrile rubber) polypropylene (EPDM) and polyurethane rubber and silicon rubber.

Fiber-reinforced plastics have a great importance to industry. Fiberglass is the preferred reinforcement, however asbestos fibers are widely used and recently "Aramids" and carbon fibers have replaced glass fibers to a small extent. Glass-fiber-reinforced unsaturated polyesters have won the largest market share. They are distinguished by high mechanical strength and are used to a great extent in building construction. They are also used in other sectors (boats, tanks) where high strength together with low weight is desired. In vehicle and aircraft construction, fiber-reinforced thermoplastics and thermosets are utilized to an increasing extent as a construction material.

6.7 Engineering Plastics

Engineering plastics are defined as thermoplastics which can be utilized as load carrying components in building construction as well as in machinery, equipment and vehicle construction. Therefore, they must have the corresponding strength and resistance against environmental influences and chemicals. The creep behavior which is observed in most plastics at normal temperatures is undesirable; dimensional stability is important. For many application purposes high temperature stability is a requisite.

The engineering thermoplastics which fulfill these requirements have a partly crystalline structure. Examples of such plastics are some nylons and polyacetals. Thermosets and specially reinforced polymers also fulfill these requirements but their processability can be more complicated than for thermoplastics.

The low density of most plastics is advantageous, however, their high thermal expansion is unfavorable being about 10 times greater as that for steel.

For the time being, these plastics are relatively expensive. Their market share out of the total consumption of plastic lies on the order of about 1%. However, their growth rate is relatively high (10–20%).

6.8 Substitution

Plastics are displacing to an increasing extent other construction materials, like metal, wood and other conventional construction material and also the commonly used packaging materials like paper, cardboard, glass, metal foil and sheet. When this process will come to a halt is difficult to say. Not only do price considerations play an important role, but also questions of special properties, suitability, esthetics and acceptability. In addition, the problem of recycling or precycling (see p. 154) has an influence. On the other hand, a substitution of plastics by cheaper or more suitable materials can take place. In the long run, each material will maintain or achieve its own special application area.

The different plastics compete among themselves in many applications. An increasing number of bulk plastics are being supplanted by more sophisticated products. A good example of this is the advance of polyurethane into numerous application sectors, displacing foamed rubber.

6.9 Investment and Production Costs

Table XXVI shows the percentage distribution for raw materials, variable and fixed costs, as well as the investment and turnover in addition to the added value. These figures are valid generally for many basic petrochemicals and for numerous plastics, despite the fact, that the many different chemical processes can give wide deviations. The largest part, as Table XXVI shows, is the raw material. Next comes the amortization and interest on the capital at 15–30%. In recent years attempts have been made to reduce this influence through the construction of larger plants. But today,

Table XXVI: Breakdown of Costs for Basic Petrochemicals

	% Turnover	% Added Value
Raw Materials	30–60	
Variable Costs	8–20	15–35
Fixed Costs	10–20	28–32
Investment Charges	15–30	40–55

for many bulk plastics, the trend to build always larger units has come to a standstill. Further cost reductions on this basis are not to be expected especially with the increase in energy and raw material prices as well as salaries and wages. In the future an increase in the price of plastics can be counted on especially as the current overcapacity slowly disappears.

6.9.1 Energy in the Plastic Industry

Large amounts of energy are used for the production of plastics. In addition, petroleum fractions especially naphtha (also natural gas) serve as raw materials. Price increases for crude oil and, as an indirect result thereof, an increase in the price of natural gas, therefore has economic consequences for the plastics industry. Plastics are therefore influenced more by these circumstances than the metals, such as iron and steel, copper as well as aluminium.

6.9.2 Environment and Pollution Control

The plastics industry, like many other basic industries especially the chemical and metal industries, has environmental problems to solve. From the view of the plastics industry there are really no unsolvable problems rather it is more a question of time and money. Of course, investments for environmental protection add to the cost of plant construction and thus result in an increase in production costs.

6.9.3 Recycling

In West Germany 6 271 million tons of plastic were produced in 1974. During the same year the amount of plastic waste generated by the raw material producers, fabricators and users (small shops and industries) amounted to 510 000 tons. Of this amount 52 000 tons were generated by the raw material producers and 96 000 tons by the fabricators. By far the largest amount of plastic waste occurs in household trash which amounts to about 1 million tons or twice as much as the amount of waste generated by industry.

A network of plastic recycling centers has been established in West Germany for a long time, these are similar to the recycling units which process plastic waste within the raw material producing industry. It is estimated that between 90–100 000 tons a year of plastic waste are processed in these units. It should be pointed out that only clean or very slightly contaminated waste is actually processed in these centers and that the different polymers like polyethylene and polystyrene cannot be mixed. Some processors sort and clean their material supply. Subsequently the waste is ground or extruded to produce granulate which has the appearance and usefulness of virgin material. Some of the recycling centers have installed fabrication equipment to produce products of differing quality. Technological possibilities are under further development so that the recycling quota can be increased.

Unusable and very dirty waste must either be buried or burned. The majority of plastic in household trash belongs to this category. The high heating value of plastics, more than 10 000 kcal/kg, predestines them for burning. Initially there was great environmental concern about burning, especially plastic waste containing PVC, since the generation of hydrochloric acid could cause serious corrosion. It has been learned that with suitable design of equipment and especially with careful temperature control any such corrosion can be avoided.

The burning of plastic waste and plastic-containing trash can provide a small but important addition to the heating and energy supply of a region. Incineration equipment should therefore be equipped for the possibility of long distance heating systems and/or energy production equipment. It is possible that in the future pyrolysis, that is, decomposition at high temperature without oxygen, can be an important way to get back some of the basic materials from which plastics are made.

The burial of plastic waste is not without problems. Most plastics are, by their nature, quite durable and their degradation under the influence of atmospheric oxygen or bacteria is correspondingly slow. The proposal is often made to develop biodegradable plastics. This presents no principal difficulty, however, plastics of this type are unusable for many purposes. Therefore, up to today, this idea has not yet been realized.

6.10 The Role of Developing Countries

Up to now, the production of plastics by developing countries plays only a minor role. For some oil producing and industrializing countries this will change shortly. There, the erection of new large production plants will contribute to world-wide overcapacity and put pressure on price levels in the future.

Other developing countries are assuming a greater role as a consumer and also to some extent as fabricators. Plastic pots, pans, dishes and tubs are preferred in developing countries and are appearing in greater numbers in the market place. Such articles are inexpensive and therefore the low-income people in developing countries can afford them. The low price for bulk plastics and the relatively small market in many small developing countries preclude the erection of capacity to fill their own needs.

6.11 Perspectives for New and Improved Technology

Compared to metals, plastics are a relatively young materials group. Some modern plastics were not even in existence ten years ago. It goes without saying that they have an enormous development potential and that their further spread into the market place is assured. The well-known thermoplastic materials have been improved like the progress in recent years has shown. Also the engineering thermoplastics, like polyamids, polymethylmetalacrylate, polycarbonate, polyacetal, polyalkyleneterephthalates and fluorocarbon plastics can undergo great further development. A few of their already good properties, which can be improved further, include toughness, resistance to electrical leakage or stray currents and resistance against solvents and crazing.

Many functional elements which are traditionally produced from metals in multiple operations such as: bearings, sliding contacts, nuts and bolts, cams, gears and other components can often be produced in one step from engineering plastics.

Thermosetting, molding-grade phenoplastics and light-colored aminoplastics have been developed which have a short time temperature resistance of several hundred °C. Dishes made from these plastics are "dishwasher-safe". Other typical applications include handles for pots, pans, irons and other uses which require heat-resistance. A very old use is in commutators for electric motors. Glass-fiber-reinforced terephtalate plastics have been approved recently for the injection molding of fuses and switches. Progress has been made with unsaturated polyester resins, where handwork previously prevailed, by low pressure pressing of impregnated resin mats. Here, further promising advances are expected and this holds true also for polyurethane.

The most important commercial process advanced during recent years was reaction injection molding of structural foams. The shape stability of polyurethane resin foams can be obtained without the loss of elasticity. The application of plastics reinforced with short-glass-fibers is advancing, also cores of glass mats with a layer of polyurethane foam is new (ski cores, coolers, aerospace equipment). Molded foams and soft foam upholstery are making further gains in the automobile industry.

6.12 Industry Outlook

Over the last three decades, the use of plastics underwent an enormous growth. Annual growth rates above 10% were observed in most of the plastic consuming countries. However, it should not be overlooked that this high growth of plastic consumption was in the field of simple products, mostly for household goods as well as other applications where structural properties were of less importance. These markets are well covered by plastics now and the growth rate has levelled off at around 5% per annum. Therefore, it would be a mistake to extrapolate the past growth of plastics into the future.

Plastics still have development potential. Therefore, in the coming years numerous new materials and material combinations will be found which, with certainty, will open up new application areas especially in the field of engineering plastics. But, the engineering plastics will never reach the huge volume attained by bulk plastics in the past, unless the automotive industry would begin using them on a large-scale for structural and/or non-suspension parts. This is an open question today that will probably not be answered until 1985. Furthermore, the development of crude oil price and availability, upon which plastics are heavily dependent, will also play an important role in their future use.

7 WOOD AND WOOD PRODUCTS

By Heinz Eldag, Mechanical Engineer and Engineer to the Wood Industry

Wood besides stone is the oldest material utilized by man. Since prehistoric times wood has been used for clubs, spears, bows and arrows as well as for shelter and heat. Wood has been available in most regions of the world. Therefore, it is deeply rooted in the entire history of civilization and mankind.

Today, wood is utilized as an engineering material in many forms, for example, sawnwood, plywood, hardboard, particleboard and veneer. Its main application sectors are housing, furniture, railroad ties (sleepers), and mining. As a matter of fact, today wood is and still remains one of the leading construction materials. Wood is an important raw material for the pulp and paper industry. Wood and wood products play a very significant role in the packaging sector. Furthermore wood is still an important fuel.

The terms softwood and hardwood are used to distinguish between the various types of timber. As an industrial material the softwoods are far more important, over 70% of wood used in industry is softwood. Over 90% of the softwood timber is located in the North Temperate Zone. The world's hardwood timber is mainly in the less developed countries in the southern hemisphere, particularly in Latin America.

The total harvest of industrial wood is about 1400 million m^3. From this amount, the pulp and paper industry uses, as raw material, a little less than 40%. When one considers that most species of wood have a density slightly less than 1, the production of wood on the basis of physical volume, is far greater than that of all metals combined. This has specific importance, because wood is the most important material available from renewable resources.

7.1 Utilization of World Forests

According to recent FAO data, the world's forest area is estimated at 4500 million hectares of which 1700 million hectares are savannah forests and bush forests of every kind.

- An intensive forest economy is performed on only 13% of the world's forest area.
- The world's present total wood utilization is estimated at 2500 million m^3/year which is less than the total utilization potential of world forests.

- Half of the wood utilized in the whole world is fuel wood, of this 80% is used in developing countries. Europe, the USA and USSR utilize 80% of the industrial wood.
- In a number of developing countries the ratio of utilization to reforestation is causing concern. There are estimates that about 12 million hectares of forests are cut down each year and not reforested, especially in Africa, Asia and Latin America. The reason for this alarming situation is the high demand for fuel wood, clear-cutting, and burning down of forests for agricultural use (mainly cash crops) and building of villages.

According to a World Bank study, all the world's forest reserves will be utilized within 60 years because of anticipated further population growth in developing countries. The United Nations Environment Situation Report 1977 foreseees an ecological catastrophy.

Viewed from a global perspective, 12 million deforested hectares each year would not appear alarming, since that is less than 0.4% of the world's forest area. By more intense planting of new forests elsewhere, this deficit could be compensated. But if one looks at local ecological niches, a rapid deforestation can result in irreversible consequences and side effects. Already in historic times the rapid deforestation of some Mediterranean regions led to soil degradation and arid areas, where it is unfeasible to grow new forests again.

Today, there are many local regions causing concern regarding negative impacts of deforestation or more intense use of forests on the ecological equilibrium. A widely publicized example is a large new pulp and paper plant combined with large plantations of cash crops in the Amazon region of northern Brazil. Here, some speculation may have entered the forecasting of negative impacts. In other parts of the world, ecological disasters caused by deforestation are already manifest. Wide areas in India still officially noted as forests are already deforested. The main reason for the growing desert in Africa is the high utilization role of fuel wood and, non-utilized waste wood from woodprocessing industries.

7.2 Consumption and Production of Wood Products

Based on data and a forecast by FAO/EEC for Europe (excluding USSR), the annual wood consumption of presently 425 million m^3 will increase up to 765 million m^3 in the year 2000. This means an increase of 2% per annum compared to a 1% increase in wood harvesting. West Germany, for example, has to import 50%. The present deficit between consumption and harvesting will increase constantly in Europe.*

An important development in the utilization of wood is taking place in the wood panel industry. While in 1949–1951, the average ratio of apparent consumption of sawnwood to that of wood based panels was 21 (21 m^3 of sawnwood used for every 1 m^3 of wood based panels) the ratios change drastically in favor of wood based panels as shown in Table XXVII.

Table XXVII:
Ratio of Apparent Consumption of Sawnwood to Wood Based Panels

Region	Ratio of Apparent Consumption 1964–66	1969–71	1974–76
Nordic Countries	6.6	5.4	4.0
EEC	5.1	3.4	2.5
Central Europe	5.7	4.0	3.4
Southern Europe	6.9	5.4	3.6
Eastern Europe	7.7	5.1	3.7
Total Europe	5.8	4.1	3.0
USSR	32.1	22.8	14.2
North America	4.6	3.7	3.3

In the USSR, the use of sawnwood still predominates over that of wood based panels (14.2 to 1 in 1974–76) but the downward trend in the ratio is likely to continue in the coming years.

The increase in the consumption of wood based panels is now slower than earlier, in many highly developed countries, because the energy consumption of wood based panels is higher than that of sawnwood. This is especially the case with particleboard. For instance the US Industrial Outlook 1979 to 1984 forecasts that particleboard will lose substantial markets to sawn timber.

The highest per capita consumption of wood based panels is in Finland (162 m^3 / 1000 capita) followed by Sweden (161), Denmark (142), Canada, USA, Norway (135 each), and West Germany (116).

The developing countries are far behind the industrialized world in the development of the wood based panel industry. While wood based panels

* There are great differences from country to country. For instance, a well planned forest economy plus geographic conditions make Austria and Scandinavian countries net exporters, whereas the EEC countries have to import wood.

from industrialized countries have a major share of exports, the developing countries have their export share in the timber trade flow still on logs and sawn lumber with a slight increase of veneer and plywood especially to developed countries. The leading importer of logs among ten countries is Japan with 6765000 m^3 followed by Italy, West Germany, France and the USA.

The leading exporters of tropical wood products are Asia Pacific, Africa and Latin America. In world timber trade, Latin American countries have not yet developed their nominal share.
The main producers of wood, in 1976 are given in Table XXVIII.

Table XXVIII:
Main Producers of Wood (1976)

Country	Million m^3
USA	348
USSR	295
Canada	136
China	53
Sweden	53
Finland	27
Brazil	27
Indonesia	25
All Others	408
Total	1 372

The developing countries will have an increasing role in the future, mainly in the supply of tropical wood. The majority of wood is produced today in planned forests in the northern hemisphere. The "harvesting" is highly mechanized and the amount of manpower has been reduced drastically. Machines are used even for reforestation. For the transport of the wood to saw mills, rivers are still used for log rafts, especially in the USA and Scandinavia.

In the southern hemisphere, in South America as well as in Africa, a trend toward a systematic forest economy can be clearly recognized. However, there is certain concern about large monocultures of wood and wood products coming into existence because they could disturb the ecological equilibrium, for instance in the Amazon area.

7.3 The Manufacture of Wood and Wood Derived Products

The growing importance of maximum yield from a tree and the lack of labor in developed countries influenced the development of new machining processes in connection with conveying and handling equipment.

7.3.1 Sawmilling

The beginning of automated production in the breakdown of logs was the linking of several conventional machines (band headrig with a resaw, an edger and a crosscut saw). To speed up the saw capacity, twin and quadruple headrigs have been developed, controlled by a TV-operator station. The increased breakdown capacity led to the development of log yard sorting and grading systems as well as lumber yard sorting and stacking systems. Depending upon the regions of coniferous or broadleaf log resources, the log conversion into lumber varies to a high extent. Since in many industrialized countries chips and sawdust are saleable products the log breakdown technology changed into chipping-sawing lines. This process starts with chipping the slabs before multi-band sawing. When using profile chippers, the main process is log profiling and bandsawing or gang sawing. With this technology logs are converted into chips and sawn lumber.

For the breakdown of tropical hardwood logs, the combination of mechanically operated log loaders, log turners, and other devices for log handling ("kickers" and "flippers") as well as board handling devices and conveying equipment have been developed to increase the production capacity.

7.3.2 Veneer and Plywood Production

Veneer is a thin layer of choice wood placed over a more common material as well as the individual layers glued together in plywood. Sliced veneer is used predominantly for decorative veneer. Peeled veneer is utilized mostly for plywood.

Rotary veneer slicing is performed on lathes and flat ("flitch") slicing on slicers which are available for horizontal, vertical or inclined versions with appropriate veneer conveying equipment.

The development of appropriate debarking machines, log loading and centering equipment was a preliminary condition to increase the peeling capacity of a veneer lathe. Sequential to this change, the appropriate automated reeling and unreeling systems, roller or chain conveying veneer dryers with jet stream drying systems were developed to speed up the manufacturing process. Automated in-line clippers with defect tracer control, veneer pack jointers and single blade jointers as well as new splicing equipment is now available for each step of automated production.

Peeler cores are remanufactured into sawn board to be utilized as core material for block boards or pallet boards. Also, in veneer production, prepeeler material or presliced material will be recycled for particle board or fiberboard production wherever possible.

The manufacture of plywood panels, plymoulds, block board, hollow core boards and flush doors are today important integrated production facilities to increase the yield of log materials.

Prepresses are developed to shorten the pressing cycle of the plywood press. Multi-daylight presses for multiple panel production and short cycling single daylight presses for single panel production, are competing with each other. In this book we cannot go into the many pros and cons in connection with adhesives to be applied and the various wood species used.

7.3.3 Particleboard Production

An important feature of particleboard production was to produce a board material for multi-purpose utilization. The particleboard production capacity of the world in 1955 was approximately one million m^3. In 1975 the capacity was 30 million m^3. The demand is still rising. The main reason for this is that particleboard may use almost any species of wood, as raw material. There are already many technical innovations used in particleboard production, for instance, strength orientation, wafering, fireproofing, and injection molding. The price level of particleboard is relatively low, which means that the product cannot be shipped long distances. It is, therefore, a typical home market product.

The general trend in the production of particleboard manufacturing equipment is towards "automation" – the computer controlled production line and development of the accuracy press. In the future, improvements of outer layers have to be developed. The final board processing is not yet appropriate to the various requirements of the industry.

7.3.4 Machining of Solid Wood and Wood Derived Products

The secondary wood processing industries are important consumers for the products of the primary wood processing industries: namely sawn lumber, plywood, particleboard, fiberboard. The machinery technology rendered possible the rise of speeds and feeds for even better and more accurate machining performance.

Today in highly developed countries, production of furniture and joinery is a real industry. There are automated lines available machining solid wooden furniture components in a 5 second cycle with up to 25 different operations and an accuracy of 0.05 mm. The West German, North American and Belgian furniture industries were the first to install this equipment. 1966 was the beginning of tape controlled woodworking machines in the USA. An American machine manufacturer at that time applied the tape control system from milling machines, for metal working, to a high speed router for woodworking. Very soon after this first introduction of tape controls, or other memory controls, various machine types, that is, panel sizing, multiboring and component sawing machines, have been equipped with computer controls. The link of single machines, forming machine groups or automated machine lines is a development applied in furniture and joinery industries. Machining lines in the furniture industry operating at flow speeds up to 40 m/min, in sequence influenced the development of the automatic readjustment of tools in double end machine lines. These machining conditions were the basis for the evolution of automatic loading, feeding, handling and stacking equipment otherwise the high machining capacity could not be utilized full time. This development would not have been successful without similar developments in the allied adhesive and lacquer industries. Adhesives and lacquers with short curing times influenced the short cycling veneering or assembling operation. Curtain, roller or spray coated lacquers are cured in hot air streams or under infrared generated heat. Also plastic joint assemblies were developed to speed up production under high accuracy.

Thus, within 30 years after the Second World War, secondary wood processing became an industry which in fact is comparable to other old established industries like the metalworking industries; a development rendered possible through the coordinating effort of allied industries forming their own infrastructure.

7.4 Wood and Competing Materials

Since wood is a renewable resource, it gains importance compared to other industrial raw materials which are under constraints of various types. A valuable feature in the provision of wood and the manufacturing of wood products, is the low consumption of energy in the sawmilling and wood working industries based on unit volume.

It is not so long ago that the plastics manufacturing industry forecasted the replacement of wood in a few years because plastic products are better and lower priced than wood. The plastic industry advertised, for example, the replacement of newsprint by a waterproof plastic material but up to that time, when the newsprinting world changes to the new process, thousands of hectares of forests have to be maintained because of the old established pulping process. But everybody knows today, that becaüse of the oil crisis, the situation plastic versus wood reversed in favor of wood.

Similar developments could be observed when the plastic industry stepped into production of furniture and similar products. We know that wood has been substituted by other materials but on the other hand, wood as a raw and/or structural material, combined with innovation, opens new fields of utilization. Furthermore, in the continuing process of substitution, wood is recently gaining momentum. It competes with steel or reinforced concrete for roof supporting members in medium sized buildings, e.g., for recreational purposes. In European countries, for window frames of private homes, wood has made a comeback and competes successfully against plastic or metals. Another area where wood is showing a high growth rate is in aluminum-wood combinations. Aluminum provides a weather resistant outer surface for example, in window frames.

Recently, tropical hardwood is a favored material for window frames, doors and premium furniture. Impregnation of wood with resins, plus high-quality finishes of all kinds have given wood additional competitive strength.

7.5 The Role of Developing Countries

Developing countries with forest resources had and have to learn that forests are economically and ecologically an important positive element if they utilize this natural resource in an enduring manner. The wood consumption in developing countries will rise more than in industrialized countries because of the huge population increase. Estimates by FAO forecast, for the year 2000, that the world demand on wood will be less than the production potential of the existing forests. But the regional unbalance will grow and the world will depend on the forest resources of the USA and USSR unless decisions are made especially in Asia, Africa and some regions in Latin America for planned forestry. Self-sufficiency in tropical forest regions, through wood and wood products at the beginning of the next century, could be achieved if for instance 150 million hectares in Asia and the Far East, 60 million hectares in Africa, and 30 million hectares in Latin America of plantations can be generated.

The self-sufficiency on wood and wood products in developing countries generates at first the establishment of small-scale industries with an intermediate technology so that these countries can develop their own wood processing industries and the adequate infrastructure. Primary and secondary wood processing industries have to be balanced in their growth, according to the conditions of the different countries.

7.5.1 Increasing Wood Yield

The development of wood processing in the future depends upon the processing costs and the price situation in comparison to the wood surplus regions of the world. It is reasonable to assume that wood prices will increase more than wages. This is true for logs, sawn lumber, semifinished products and finished products. Exporting developing countries and some industrialized countries know already that the present prices are insufficient to activate reforestation on exploited areas which would contribute to a healthy forest policy. Attempts by developing countries to stabilize and increase wood prices within a world-wide framework are already in action. The industrialized world has to assist the developing countries especially in their forest policy but more than ever in establishing an appropriate wood processing industry with increased yield of the raw material wood. Industries in developing countries achieve, on an average, 23% yield from log input compared to a 65% yield from log input by industries in developed countries. This demonstrates the losses which could be utilized for wood products or as an energy source for generating heat and electric power. The wood gas process, by using modern gas motors, is one of the most energy-efficient processes, an old process combined with new ideas*.

7.5.2 Outlook on World Forestry

The natural resource of wood, its balance in reforestation and harvesting, the possibility of increased growth which will guarantee the availability of wood in the future, needs the following conditions:

1) increased forest growth especially in developing countries;
2) evaluation of ecological aspects on the basis of profitability (cost-benefit-risk-analysis) and opposition to unjustified environmental demands;
3) a free market economy which also requires a liberal foreign trade policy for wood products.

Developing countries should not consider export markets only. They have to build up their own industry in using as much of their forest resources as possible to avoid imports.

* In fact, the most economical system to provide an isolated village in an underdeveloped country with electricity (a few hundred watts per house) is the operation of an electric generator with a simple gas engine, the gas being provided from pyrolysis of wood or other agro waste.

Research on secondary wood species is a preliminary condition to open new markets and new fields of application. That means, industrialized countries have to reconsider their wood markets in introducing new species or semifinished products which they wish to import.

7.6 Use of Wood for the Production of Chemicals, Pulp and Paper*

The major emphasis in this book is on engineering materials, so we will limit ourselves to a few remarks about wood use by the chemical industry.

The importance of wood as a feed stock for the chemical industry has decreased very much over the last decades. Previously it was used on a larger scale to produce charcoal, acetic acid, methanol, tar and similar products. Many other products were obtained from the tar. Dissolving pulp for producing chemicals , fibers and films (cellophane) represents today only 3% of total wood pulp (see Table XXIX). A few years ago it appeared that viscose fibers would continue losing ground to synthetic fibers. However, increases in oil price have at least stabilized that situation.

7.6.1 Pulp and Paper

The pulp and paper industry is a large consumer of wood. The production is carried out in highly specialized plants which use different processes, depending mainly on the type of wood and application (sulphite process, sulphate and soda process, mechanical grinding, semi-chemical and chemiground wood process).

Table XXIX shows that the pulp industry has a great economic importance. Its production exceeds that of the non-ferrous metal industry by a wide margin. The pulp industry is based predominantly on wood as a raw material. Today, other raw materials have only an unimportant share. The recycling of old paper and paperboard already is a significant factor and it will increase with time. It is estimated that pulp production requires about 200 million tons of wood per year. This makes the pulp industry one of the largest wood consumers. Nevertheless wood consumption is confined to a few species such as: pine, spruce, poplar, eucalyptus and in smaller amounts, birch, beech and some others coming mainly from planned forests.

* contribution by Dr. G. Friese, Swiss Aluminium Ltd.

Table XXIX:
Total Woodpulp Production in OECD Countries in 1000 mt (1977)

Sulphate and soda woodpulp	56 522.9		
Sulphite woodpulp	7 866.7		
Total chemical pulp		64 389.6	
Semi-chemical and chemi-ground wood pulp		6 565.4	
Mechanical pulp		20 780.1	
Woodpulp for paper and paperboard making			91 735.1
Dissolving pulp *			2 923.8
Total woodpulp			94 658.9

* Mostly for the production of artificial fibers and chemical use
Source: OECD, "The Pulp and Paper Industry 1977/78", Paris.

By far, the major part of pulp production today is used in the manufacture of paper and paperboard (> 95%). In 1977, paper and paperboard production in OECD countries was more than 120 million tons. It is one of the largest branches of industry. Paper still plays a dominant role as a writing and printing material as well as in the packaging sector and other important areas. This will not change much in the future and, increases in energy costs together with possible energy shortages could be favorable for the paper industry.

Leading countries in the pulp and paper industry are the USA, Canada, Japan, Sweden and Finland as well as a few other European countries. Outside the OECD, the USSR is an important pulp and paper producer with huge forest reserves, the largest part of which is in untapped Siberia.

For the time being, the developing countries do not figure largely in the pulp and paper industry, but their importance will grow quickly (e.g., Brazil).

7.6.2 Environment and Pollution Control

The harvesting of wood and the production of engineering materials from wood presents no difficult environmental problems, which also holds true for workers hygiene. However, this is not the case for the pulp and paper plants whose main problems are water pollution and air pollution. A continuing problem with the sulphate plants is annoying odor.

Great efforts are being made to improve the situation at pulp and paper plants. Scandinavia is a precursor, but the USA and even Brazil are now in the process of a very intense clean-up of the waste water stream from the paper and pulp industries mainly by concentrating and burning the spent liquor, which covers a great part of the heat and energy demand of the plants.

7.6.3 Recycling of Paper and Paperboard

Of the total paper and paperboard consumed throughout the world, about 75% is made from virgin fiber. The remainder comes from waste paper and other materials which have been recycled. These proportions have not changed much in recent years, but there is increasing interest in recycling. In many countries, studies are being made as to the possibility of recycling larger quantities of paper and paperboard. In some countries in Western Europe such as West Germany and Britain, where sources of fiber are small in relation to needs, waste paper provided more than 40% of the fiber required for paper and paperboard production. In Scandinavia and North America, where forest resources are more plentiful and most pulp and paper mills are located closer to forests than to large urban centers where substantial quantities of waste paper are available, the percentage recycled is somewhat smaller. In the United States, the proportion of secondary fiber to virgin fiber is around 20%. However, this is expected to increase as techniques for collecting, transporting, and using waste paper as a raw material are improved.

The basic pulp used in the manufacture of newsprint is mechanical pulp made by grinding wood against a stone or by refining chips to pulp in a thermomechanical or chip-refiner system. The energy consumption per ton is higher than that for chemical pulps used in container board. Thus increased use of recycled newspapers in making newsprint reduces overall energy usage by industry as well as pollution loads. In addition to reduced energy requirements through recycling, a major benefit is the alleviation of the solid waste problem facing major metropolitan areas. In the USA, waste paper constitutes 30 to 40% by weight of total municipal solid waste and about 15% of the waste paper is newsprint. In order to use old newspapers they must be de-inked before they can be reformed into new paper. The main constraint in broader application of de-inking old newspaper for newsprint manufacture is the number of metropolitan areas with a sufficiently large population to generate the volume of old newspapers needed to support a recycling operation.

7.6.4 Precycling

Besides recycling, wood products and paper lend themselves to precycling. Precycling means previous use of a material before it is burned to recover its heating value (plastic packaging is another example of precycling). Modern plants generating energy by burning municipal waste are the instruments for precycling.

7.6.5 Composting

In forests, bark and wood waste can be spread around to serve as a soil conditioner. Also, recently industrial organic fertilizers have been produced from wood and bark residues. Therefore from the standpoint of recycling or precycling as well as utilization of production waste, wood is an excellent material.

7.7 Perspectives for Improved Properties and Products

Wood's properties are rather anisotropic and they vary from tree to tree, as well as within the same tree trunk. This is largely overcome in plywood and wood panels by suitable technology.

Despite the fact that wood is the oldest material utilized by man, its development has not come to a standstill. In recent years, numerous new developments have taken place which have broadened the application of wood and wood products. Especially noteworthy are new impregnation techniques and combinations with other materials such as plastics and metals, which will help assure the use of wood in the future. Wood's main advantages lie in its relatively low price, favorable strength-to-weight ratio and the ease in which it is worked with proven techniques.

One of the more fascinating areas for innovation is the drive for better utilization of wood waste. It must be remembered that in felling, and processing wood, the outer parts of the original tree are cut away and sometimes abandoned. This material and the by-products from saw mills and paper mills, that is, sawdust and bark, can be used to produce either pelletized clean fuel or provide a material for the production of composite products which have wood, plastic, or metal as an outer surface.

In conclusion, it can be stated that wood products and wood technology have a bright future.

8 ADVANCED MATERIALS

The so-called "materials revolution" gathered momentum after World War II. Significant achievements in advanced materials and corresponding materials systems occured in the last three decades. In the United States very advanced materials were developed within the space program. These materials combined either extreme lightweight with high strength and/or resistance to high temperatures. They were developed by the National Aeronautics and Space Administration (NASA) and its subcontractors. During the late sixties and in the seventies, numerous "NASA spin-offs" emerged to transfer these sophisticated materials into the private sector. Simultaneously in Europe and Japan, equivalent developments took place. These innovations were carried out by research institutes and quite different materials industries which produce plastics, metals, ceramics and all kinds of composites.

8.1 Three Main Target Areas for Advanced Materials

A. Lightweight but strong materials: This materials class is now in focus for many civilian applications such as: aircraft and road or rail vehicles. Included in these materials are metals like aluminium, magnesium and titanium; new "exotic" plastics for engineering applications, and fibers consisting of various organic or inorganic substances.

B. High temperature materials: Gas turbines are the main target area here, since raising the operating temperature above 1200°C would yield a significant improvement in efficiency. Such an achievement would even open up new perspectives utilizing smaller gas turbines for road vehicles or railroad use. This is a typical example where engineering development is very much dependent on the provision of advanced materials.

C. Materials for electronic components: Examples of end uses are: solar photovoltaic cells, rectifiers and transistors to name a few. The materials are often single crystals of silicon, germanium or gallium compounds. Thin films or multiple layers of semiconductors already play a major role in microcircuitry and this development will continue.

8.2 Typical Examples of Ongoing Innovations

8.2.1 Fiber Reinforced Materials (Composites)

Composite materials generally consist of a matrix material through which a different, reinforcing material is distributed. Fiber composites originated with glass-fiber-reinforced plastics. The fibers have high strength, but relatively low stiffness. Glass-fiber-reinforced plastics have long been used in structural applications where a high strength-to-weight ratio is the predominant requirement. Furthermore improved glass fibers came on the market which could even be embedded in concrete instead of conventional reinforcements (without embrittlement because of special alkali resistant glass fibers).

In the sixties, fibers of boron, graphite, silicon carbide, and other materials with high stiffness were developed. These newer fibers are used in both resin and metal matrices.

Initially, the highest strength fibers like graphite and boron were limited to space and military applications, because of cost. Today, high-performance fiber composites are already used in aircraft. They are being considered as a replacement for metals in automobiles, where they could perform the same functions with a 50–70% weight saving. However, despite a sharp drop in the price of graphite fibers they are still considered too expensive to be used widely in the automotive industry. Therefore, much effort will be made during the next five years or so to find new resin formulations and cheaper ways to produce fibers and to develop manufacturing methods for high-volume, low-cost composite parts.

But there is no question that reinforced plastics can go a long way. The "materials revolution" in this field is just at the beginning of the second strata of substitution where tailor-made plastics with improved properties replace bulk plastics used previously.

It is exciting to see how within two decades fiber reinforced resins, in a first level of substitution, have replaced, e.g., metals, mainly steel, partly aluminium in the container field. Today up to 90% of all medium or large containers for liquids or other bulk materials with volumes up to 1000 m^3 are made from fiber reinforced plastics.

8.2.2 Ceramics in Competition with Metals

Superalloys are alloys intended for service at temperatures above 750°C. They are used routinely in jet engines. Higher operating temperatures of around 950°C, which improve the performance and increase the efficiency of jet engines as well as reduce fuel consumption, were achieved by special microstructures and oxidation resistant coatings. A further development has been the use of directional solidification in nickel-base superalloys to produce turbine blades. However, despite all these advances, as shown in Figure 17, the existing superalloys have their limits at about 1100°C in gas turbines. In continuous use at higher temperatures they fail due to oxidation, erosion or creep.

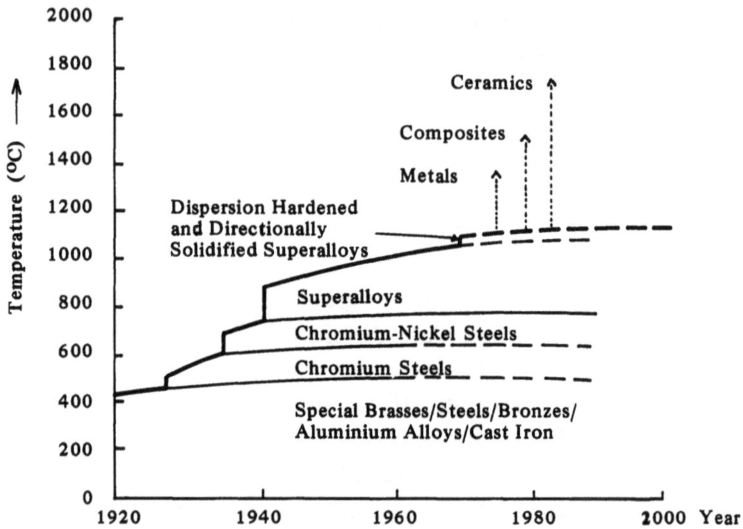

Figure 17: Chronological Development of High Temperature Materials
Source: G. Petzow

At temperatures well above 1200°C, ceramics would be the ideal substitute for metals. There are well advanced developments to produce ceramics with a ductility similar to cemented carbides now used for tools. The typical test case for high temperature use of ceramics are turbine blades. There are simultaneous developments to improve metallic systems and composite materials somewhat further. But it looks like – à la longue – the ceramics will have a certain domain at elevated temperatures, where the metals cannot compete.

8.2.3 Advanced Use of Metals and Alloys

This is a broad field and we can only make a few remarks indicating trends and possibilities. Even well-known metals like titanium and magnesium could enter into wider use as "advanced materials" as soon as price is competitive enough. Because of the need for lightweight metallic components in aircraft and automobiles and other transport vehicles, there is a high probability that titanium and magnesium alloys and composites will have a rapidly growing market during the next two decades.

Titanium use was hampered not only by price, but also by difficulties in "users technology". However, today titanium is entrenched as an aircraft structural material and has a growing market in chemical plants as a corrosion resistant material.

Magnesium up to now has been used in non-alloyed compositions or in a few rather simple alloy systems. But if magnesium would enter a stage of more intense development, into an advanced material, there are lots of opportunities to improve its corrosion resistance and its strength at normal or elevated temperatures. In addition, magnesium could find use in the form of castings for the transportation industry.

We chose these examples to explain that well-known materials, by entering the second order of substitution (by adding more know-how or advanced technology to the previously used bulk material), could have a very interesting future. The same, by the way, holds true for many conventional materials, even wood, cement, steel or aluminium, to name just a few examples. Returning to the advanced use of metals, we wish to restrict ourselves to just two developments pointing out the importance of materials science as a source for innovation.

In the case of steel, a whole family of medium strength steels called microalloyed or dual structure steels came into larger use. At the same time, in Japan and elsewhere, developments are well advanced to make steel on an experimental scale, with a strength of several thousand Newtons/mm^2, approaching perhaps half of the strength of an iron "whisker" which has very few lattice defects.

It is difficult to predict at what time in the future these advanced steels could replace today's steels on a larger scale. There are indications that this may be the case about 20 years from now.

The last example is glassy metal. Normally, metals have a crystalline structure. However, certain alloys can be solidified in non-crystalline, or amorphous, form, like glass, using very fast cooling rates (100 000°C–1 000000° C per second). These materials consist of metals like iron, cobalt, and nickel, alloyed with elements such as phosphorus, silicon, and boron. They can be solidified in a glassy form as ribbons.

Glassy metal magnets have high mechanical strength and improved magnetic properties. This makes them good candidates for replacing silicon steel sheets used in transformer cores by increasing efficiency and saving energy. This is just *one* of many examples how advanced materials are important in the provision and use of energy. Further examples are cited beginning on page 197.

Our final remark points to the way how improved or totally new alloy systems have been detected so far. The method was mostly experimental, using "trial and error" to combine different metals with each other. Considering the 80 metals known up to now, one can imagine how many combinations are possible. Table XXX shows that by combining three metals (ternary-alloys) 82 160 possible alloy systems would exist and in each system many different compositions could be used. By combining 6 metals, more than 300 million alloy systems are possible, again with many different compositions within one system.

Table XXX: Systems Possible with Various Components

Number of Components	Number of Possible Systems
1	80
2	3 160
3	82 160
4	1 581 580
5	24 040 016
6	300 500 200
7	3 176 716 400

Source: G. Petzow, Max-Planck-Institut, Stuttgart

In recent years, the understanding of structural features of alloys in combination with their properties has made great progress. Now it seems feasible to predict, with computer models, the constitution of alloys and some correlation to properties of new "exotic" alloys containing several and perhaps up to ten different metallic elements.

All this is still in a somewhat early stage, but it represents an enormous challenge not only to materials science, but also to materials producing and using industries in highly developed countries.

8.3 Conclusion and Outlook

It is impossible, in a short book like ours, to go into the many different aspects of advanced materials. However, to provide a birdseye introduction into changes in the materials industry, they can be crucial elements in future perspectives. It may be even a matter of survival for certain materials industries, in highly developed countries, to concentrate more and more on the development of "intelligent materials" which can be introduced by replacing the more simple materials like the previously used ordinary steels, plastics or other traditional materials.

Whereas in the last three decades, spin-offs from the aerospace industry were often the driving force ("science push") to introduce advanced materials, in the future it may be more a market pull, originated by the need to save energy or to conserve resources. It is, however, in a way tragic, to see that large basic material industries, in Europe and the USA, are still very much a "captive" of their previous product mix by not having enough innovation ready to enter the world of advanced materials.

9 WHERE ARE THE BASIC MATERIALS INDUSTRIES HEADING?

In this chapter we have discussed the principle basic materials industries, upon which modern civilization is based. We have shown that material resources, to become available for man's use, must go through complicated processes. Man himself plays a significant role because without his ingenuity, knowledge and technological capabilities there would be no usable industrial products. In addition, man determines the demand for materials through his demand for products. This demand for materials has given rise to questions about the future availability of basic materials. For this reason the doomsday prophets get attention from time to time. However, historic projections of scarcity have proven to be wrong. Nevertheless the doomsday prophets are worth listening to, if they stimulate thinking and the action necessary to prove their projections wrong.

Of the basic materials we have covered, only wood is a renewable resource. But this does not mean that we can have all the wood we want, when we want it. It takes time to grow trees.

Time is a factor for every industry. Lead time, that is the time between the initiation of a request and its fulfillment is a reason why additions to supply cannot be effected in a few years. All basic materials industries must resolve a number of geological, technical, ecological, financial and political questions before they start to dig new pits or open mines, or erect new smelters and refineries.

Although we have distinguished between industrialized countries, the Eastern Bloc countries, and developing countries, the basic materials industries in all of these countries must resolve the following questions before and during planning additional capacity:

- where is the natural resource located,
- by what method can it be extracted,
- how and where will it be processed,
- what are the impacts on natural environment,
- what will it cost, and
- what is its commercial value at current and likely future prices?

In addition, there are questions regarding government involvement such as financial support, environmental control, policy issues and international relationships. All of these combined will determine economic feasibility plus social and political acceptance.

On the supply side, there is no foreseeable physical shortage of basic materials. But there are other limits which may arise, mainly temporary bottlenecks in production capacity from the mine to the smelter and possible local constraints in energy supply to the materials industry.

On the demand side, the fact that the contribution to GNP by materials, in industrialized countries, is declining should not be misinterpreted, because per capita consumption is increasing further for all important materials.

Therefore, the basic materials industries face the task of supplying increased demand. Due to economic cycles, periods of surplus may occur. Various world bodies have as an aim, an increased standard of living for an increasing population. This means that there will be an expanding demand for materials. How to cope with this demand is the problem. In the past there has been too little planning, bad or insufficient data and a belief that statistics and computer programs could serve as a policy base rather than good sense and good data.

Innovation will remain a key issue for most of the materials industries and it will encompass the entire materials cycle. It is interesting to compare where the gravity center of innovative efforts has been in specific industries up to now. The steel industry's drive for innovation was mainly in the steelmaking processes whereas the aluminium industry pushed forward mainly in tailor-made alloys and many new uses for the metal. This may partially invert because new microalloyed steels are being developed for specific end uses and the aluminium industry is making great efforts now to arrive at new or improved extraction and smelting processes.

The rather young materials group which includes the light metals, engineering plastics, composites and high temperature materials have arrived at high engineering performance using the results of materials science.

This interdependence works both ways and will be the backbone of many new developments in materials use, from theory and concepts through testing and design into new or improved products.

In the following chapters we will discuss the importance of innovation and planning, possible constraints and mechanisms such as substitution and recycling to enable material supplies to be extended.

Bibliography

R. *Albin*, "Die Bedeutung technischer Kriterien für die Planung von Betrieben der mechanischen Holzindustrie in Entwicklungsländern", Holztechnologie Nr. 23, Sept. 1978

D. *Altenpohl*, "Aluminium Viewed from Within", Aluminium-Verlag, Düsseldorf, 1980

The *Aluminum* Association Inc., "Aluminum Statistical Review, 1978", Washington, D.C., 1979

American Association for the Advancement of Science, "Science" Vol 191, No 4227, Feb. 20, 1976

M. *Berthier* and M. Dangeard, "Trends in Capital and Operating Costs in the Petrochemical Industry and the Effect on Profitability", Compagnie de Petrochemie / Societe Nationale des Petroles d'Aquitaine, Paris 1971

Prof. Dr. L. von *Bogdandy* and Prof. Dr. H.-J. Engell, "Die Reduktion der Eisenerze", Verlag Stahleisen 1967, (engl. version in 1971)

Cembureau, "Annual Reports of the Cembureau", Brussels, 1973/1978

*Cipec** "Map of World Copper Consumption", Neuilly sur Seine, France, 1975

Cipec "Statistical Bulletin, 1977", Neuilly sur Seine, France, 1978

J. Kenneth *Craver* and Roy W. Tess, "Applied Polymer Science", Organic Coatings and Plastics Chemistry Division of the American Chemical Society, Washington, D.C., May 1975

W. *Dettmering*, "Eisen im System unserer Welt", Stahl und Eisen, Nr. 25, 1975

Hans *Domínghaus*, "Kunststoffe", VDI-Verlag Düsseldorf, 1969/1972

Hans *Domínghaus*, "Verbesserte bekannte und neue Kunststoffe als technische Werkstoffe", Synthetic, 1978

M.F. *Dowding*, "The World of Metals", Metals and Materials, July 1978

H.M. *Drink* and H. Rühmann, "Erfassung von Kunststoffabfällen", Forschungsprogramm Wiederverwertung von Kunststoffabfällen, Institut für Kunststoffverarbeitung in Industrie und Handwerk an der Rheinisch-Westfälischen Technischen Hochschule Aachen, 1977

Environmental Committee Report, "Fluoride Emissions Control: Costs for New Aluminium Reduction Plants", International Primary Aluminium Institute, London, 1975

Food and Agriculture Organization of the United Nations, "Yearbook of Forest Products", 1976

Werner *Gocht*, "Handbuch der Metallmärkte", Springer-Verlag 1974

A. *Hurlich*, "An Overview of Materials Shortages", ASM News, Metals Park, Ohio, August 1976

International Iron & Steel Institute, "World Steel in Figures 1979", IISI, Brussels, 1978

K. *Kehr*, "Evaluation of Wood Residues as an Energy Source for Forest Industries", UNIDO Document ID/WG 296, August 1979

"*Knaurs* Weltspiegel 1979", Droemer, Knaur

Wassily *Leontief*, Ann P. Carter, Peter A. Petri, "The Future of the World Economy", Oxford University Press, New York, 1977

Arthur D. *Little* Inc., "Economic Impact of Environmental Regulation on the United States Copper Industry", January 1978

D.H. *Meadows* et al, "The Limits to Growth", Universal Books, New York, 1972

Prof. Dr. O. H. C. *Messner*, "Metall-Interview", Metall No 11, November 1978

Prof. Dr. O. H. C. *Messner*, "Wie steht es um unsere Materialversorgung?", Schweizerische Spenglermeister- und Installateur-Zeitung, Nr. 19, 9 Sept. 1977

Metallgesellschaft A G, "Metallstatistik", since 1893, Frankfurt am Main

Dr. E. *Michaelis*, "8. Internationale LD Arbeitstagung / Eröffnungsrede", June 1977

W. *Michalski*, "Industrial Raw Materials", The OECD Observer, No 93, July 1978

* Intergovernmental Council of Copper Exporting Countries

National Commission on Supplies and Shortages, "Government and the Nations Resources", US Government Printing Office, 1976

Nato Science Committee Study Group, "Rational Use of Potentially Scarce Metals", NATO Scientific Affairs Division, Brussels, 1976

David *Novick*, "A World of Scarcities", Associated Business Programmes Ltd. London, 1976

G. *Petzow* and Hans Leo Lukas, "Konstitution, Fortschritte und Tendenzen", Zeitschrift für Metallkunde, Heft 12, Band 61, 1970

G. *Petzow*, "Hochtemperaturwerkstoffe", Zeitschrift für Werkstofftechnik / Journal of Materials Technology, Heft 4, 1972

Prof. Dr. E. *Plöckinger* and Dr. O. Etterich, "Elektrostahlerzeugung", Verlag Stahleisen 1979

N. *Rehbock*, "Die zukünftige Bedeutung des Rohstoffes Holz in Industrie- und Entwicklungsländern", BMW, Bonn, May 1979

Schriftenreihe des Fonds der Chemischen Industrie, Heft 14, Frankfurt am Main, 1978

Dr. W. *Sies*, "Afrika als Teil der Weltrohstoffwirtschaft", June 1978

Dr. W. *Sies*, "Probleme der Weltrohstoffwirtschaft am Beispiel der NE-Metalle", May 1978

J. *Szekely*, "The Future of World's Steel Industry", Marcel Dekker Inc., New York, 1976

J. *Szekely*, "Toward Radical Changes in Steelmaking", Technology Review, February 1979

UN/EEC Document, "Medium-term Survey of the Wood-Based Panels Sector", Geneva, February 1978

US Bureau of Mines, Mineral Commodity Profiles, "Aluminium" MCP-14, Pittsburgh, 1978

US Bureau of Mines, Mineral Commodity Profiles, "Copper" MCP-3, Pittsburgh, 1977

US Bureau of Mines, Mineral Commodity Profiles, "Iron Ore" MCP-13, Pittsburgh, 1978

US Bureau of Mines, Mineral Commodity Profiles, "Iron and Steel" MCP 15, Pittsburgh, 1978

US Bureau of Mines, "Mineral Trends and Forecasts", Washington, D.C., 1979

US Bureau of Mines, "Minerals Yearbook", Washington, D.C.

US Department of Commerce, "Proceedings – Materials and National Issues Conference", Washington, D.C., November 18, 1977

US Department of the Interior, "United States Mineral Resources – Geological Survey Professional Paper 820", US Government Printing Office, Washington, D.C., 1973

Verein Deutscher Eisenhüttenleute, "Gemeinfassliche Darstellung des Eisenhüttenwesens", 17. Auflage 1971, Verlag Stahleisen

Dr. J. Th. *Wasmuht*, "Das Stranggiessen von Stahl", Stahleisen-Schriftreihe 1975

World Wood, Vol. 19, No 6, May 31, 1978, Mella Freeman Publications, San Francisco

III Technology Planning as Part of Industry's Planning Process

1 Why Technology Planning?

Chapter II has given an overview of the basic materials industry and its growth during the first three-quarters of this century. Main criteria for the development of these industries in the past were achievement of growth rates in production and consumption as high as possible, and technological improvements for further increasing output per production unit.

But since the early 70's new criteria have entered the decision-making process for technology planning.

In developing countries, the demand for appropriate technology is clearly recognized, as well as processing raw materials within the country to add as much value as possible. If, for resource-rich developing countries, this concept is to be carried out with consequence, but also with careful understanding of the market forces, tailor-made solutions must be found for each of the basic materials and for each specific case.

In highly developed countries, three fundamental changes are to be observed.

- For basic materials industries, growth of output in tons is not a desirable main criterion any more.
- A steady but accelerated trend towards adding more know-how in the use of materials and their properties.
- Last, but not least, the demand by people for better quality of life will be a main driving force in industry's activities. This term, "quality of life", comprises not only environmental protection, but in general, providing new and more sophisticated products and services.

In the future, it will not be desirable and often not even possible, to perform activities related to materials and energy which are technically possible, but which are not feasible from the viewpoint of society. This relates to both existing and proposed new plants.

Looking at these basic changes in developing and industrialized countries as well, it is clear that adapted industrial activity cannot be expected to come by itself, or just through market forces or science push. To arrive at the appropriate technologies, systematic technology planning is an important task. In this chapter, we will briefly describe the main criteria and sequence of technology planning.

2 Technology Assessment as a Main Element of Technology Planning

Technology assessment is the analysis of actual or potential impacts of a specific technology or product on social, political, economic and/or environmental values. It goes far beyond assessing the technology itself – its main concern is how technology will affect nature and people in general. This is specifically necessary for unintended long-range and/or indirect impacts.

In the early 70's, the US Congress decided to set up an Office of Technology Assessment (OTA) to advise the US Congress on important technological issues. The main issue was, whether or not an in-depth assessment of the total cost-benefit and risk of a specific technology or product would yield agreeable or unbearable risks or consequences.

A whole methodology for the execution of technology assessment (TA) was worked out. The main goal of TA was declared to help prevent conflicts between government, industry and the public. Such conflicts had been very numerous and since the early 60's created enormous concern to the public and mass media in most of the highly developed countries.

The basic industries soon ended up in the center of these worries. The interesting book "Technology and Social Shock" by E.W. Lawless, describes about 50 cases where a specific technology or product created great concern and where over the years public participatory assessment started a certain pressure until government and/or industry took stronger action. Most of the cases were related to the Chemical and Food industry. But there are also numerous cases related to the materials industry like the Donora air pollution case of 1948. There, in a heavily industrialized valley in Pennsylvania; with steel, wire and zinc industries, 43% of the population suffered respiratory disease (6000 cases) from air pollution within a weeks time, and approximately 20 people died. Other cases relevant to the materials industry, for carrying out a technology assessment, are the asbestos health threat from construction and insulating material containing asbestos or the tailings from taconite beneficiation, which contaminated drinking water taken from Lake Superior in the USA.

The previously quoted book by E.W. Lawless gives an uniform analysis of all these case histories by using the methodology of technology assessment. However, the 50 examples from this book are all retrospect TA's. But the real virtue of technology assessment is anticipatory, as part of the materials industry planning process.

Industry was rather reluctant to accept technology assessments and particularly the activities of OTA at the beginning. Especially in the chemical and materials industry, the fear emerged that technology assessment might change very easily into "Technology Arrestment" or "Technology Harrassment". To avoid this, we suggest that industry itself should become engaged in technology assessment as one element of its technology planning. Most planning of technological activities is actively carried out in industry and not by governments or other institutions.

3 Main Criteria and Sequence of Technology Planning

Figure 18 describes the main criteria and sequence of technology planning for the materials industry. As this figure indicates, at every stage of the planning process there are possible feed-back loops to previous stages. To proceed according to this sequence is not only useful for the materials in-

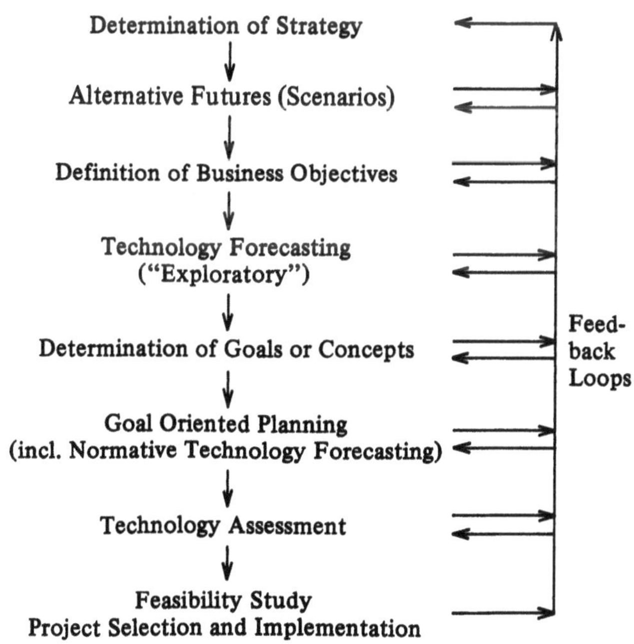

Figure 18: The Sequence of Technology Planning

dustry, but certainly for regional or national planning related to natural resources as well.

This planning scheme sets up a strictly logical sequence. It begins with strategy determination like answering the question "In what business do we want to be by 1990?".

To answer this question, different assumptions are made to develop scenarios (normally 2 or 3, sometimes up to 5) to provide a basis for determining medium to long-range business objectives. Then, new or proven technologies fitting these objectives are investigated in the next step.

This leads to specific goals or concepts, fulfilling all conditions and opportunities, determined in the previous steps of technology planning. Goal oriented planning now leads to the important step of technology assessment where all impacts of the chosen technology or product have to be carefully analyzed. This procedure provides important inputs for the process of conventional planning and implementation which is based mainly on financial criteria.

We will now show how technology assessment fits into the sequence of technology planning, using an aluminium producer as an example.

4 Technology Planning Using an Aluminium Producer as an Example

1) Determination of Strategy

An aluminium producer wants to strengthen his position in the aluminium market (primary and secondary metal)

2) Alternate Futures (Scenarios)

- No new electrolytic smelters in highly industrialized countries
- New conventional smelters in developing countries possible
- Recycling will be a main issue

3) Definition of Business Objectives

To maintain profitability, innovation and growth of trend conforming activities are prerequisites.

Aluminium can grow further, if the industries producing and using aluminium can adapt to the availability and price of energy, and if the recycling rate increases.

4) Technology Forecasting ("Exploratory")

- What processes, systems or products, that use less energy, will be available?

- What kind of substitution is likely to take place?
- How can more know-how be added to aluminium, what systems make functional use of the metals specific properties?

5) Determination of Goals or Concepts

- Find new smelter sites in "energy islands" and under acceptable conditions.
- Develop processes, using at least 20–30% less electricity for smelting aluminium than the conventional electrolysis.
- Develop direct reduction of aluminium from ore by thermal energy, even if the produced aluminium has some impurities like silicon.
- Provide or develop technology or systems to facility recycling of used aluminium products like cans or car components.

6) Goal Oriented Planning (Including Normative Technology Forecasting)

Background for planning:
- Which of the conventional and/or new processes, described under step 4 will meet the requirements of step 5?
- What will be their technical status by 1985 or 1990?

Planning for a given company and/or country:
- Main incentives or constraints of specific technologies.
- What is unresolved, so far?
- What must be done to achieve success?
- Determination of specific action targets.

7) Technology Assessment

Is the proposed technology and specific target compatible with assumed energy constraints?

Will there be other side effects, if so, what will be their influence on the total eco-system?

8) Conventional Planning and Implementation

Feasibility Study
- What will be the cost-benefit and risks for the main alternatives?

Project Selection
- Which alternative will best meet the company's innovative, marketing, financial, engineering capabilities etc.?

Implementation
- Feed-back loops to all previous steps have to be kept in mind because of the amount of time which elapses during the entire planning cycle.

Such systematic planning, in 8 steps, is worthwhile if major decisions and different options are at stake. In practice, one key element will often dominate the technology planning, for instance the availability of an important new process.

To quote a recent example: It is now possible, in continuous operation, to convert aluminium alloys from liquid metal into wide strip within a few minutes time by new or basically improved strip casting processes. This by-passes expensive scalping, soaking and hot rolling of slabs and therefore reduces investment, labor and energy for the production of sheet and coil. Such a quantum jump immediately enters the technology planning by aluminium producers and users as well. One of the reasons: strip casting could be instrumental in establishing decentralized recycling of scrap like used cans or automobile body sheet.

A final remark refers to the step of technology assessment. This is often carried out as a separate investigation, standing on its own merits. Specific assessments are often required today, like an "environmental impact statement" or an "energy audit" for existing or new production units.

5 Importance of Technology Planning for the Materials Industry

In the last subchapter, we have chosen for demonstration purposes one specific example – the aluminium industry – to explain in a very abbreviated style, how technology planning should be carried out in a materials industry. For the energy industry or companies, where materials, energy and/or ecology are major elements of planning future activities, it is of utmost importance to put strong emphasis on systematic technology planning. Further, to bring this up-to-date at regular intervals and/or when a fundamental change takes place, for instance in resources, technology and/or market pull relevant to the company's activities.

Let us briefly consider the automotive industry as another example. This materials-intensive industry is faced with a radically changing total environment. In the past, technology planning by the automotive producers was mainly oriented towards cost reduction and better performance. Now, technology assessment, within the framework of technology planning, deserves increased emphasis. When looking at the 8 planning steps in Figure 18, one can imagine that already in step 1 – Determination of Strategy – as well as in all subsequent steps, important and even dramatic alternatives will emerge.

The automotive industry is not only producing cars which are more fuel efficient, but it is also reducing the energy required for each car produced.

This brings a host of factors into the planning and decision-making processes. Some of these are:
– other fuels like gasohol and hydrogen;
– modified or new engines;
– battery power systems;
– materials substitution.
On top of these and many other considerations, relevant social elements must be recognized and taken into account. It is not easy to determine what people really want and what they will buy. But, further insight could be provided by a thorough technology assessment. In Chapter IV, we will outline how some of the substitution will be initiated within the automotive industry from the social arena, or by energy constraints or new technology (see pages 148, 161 and 172).

Determination of strategies or aims becomes more complex when the technology planning process is applied to a nation or a region. Then it enters the socio-political arena even more and quite divergent interests may emerge.

Today and in the future, for any new endeavor, the principle of well-selected growth targets and therefore, well-chosen technology, assumes increased importance. This is specifically true for the whole life cycle of materials.

Instead of describing other cases of how technology planning is carried out in a specific materials industry, we would rather point out key issues for technology planning and technology assessment related to materials.

Bibliography

D. *Altenpohl,* "TP Die Zukunftsformel" Umschau Verlag, Frankfurt am Main, 1975

Robert U. *Ayres,* "Technological Forecasting and Long-Range Planning", McGraw-Hill Book Company, New York, 1969

James R. *Bright,* Milton E.F. Schoeman, "A Guide to Practical Technological Forecasting", Prentice-Hall, Inc., Englewood Cliffs, N.J., 1973

Marvin J. *Cetron* and Christine A. Ralph, "Industrial Applications of Technological Forecasting" – "Its Utilization in R & D Management", Wiley-Interscience, a Division of John Wiley & Sons, Inc., New York, 1971

Marvin J. *Cetron,* "Technological Forecasting – A Practical Approach", Gordon and Breach, Science Publishers, New York, 1969

E.Q. *Daddario,* "Keynote Address", Conference on Requirements for Fulfilling a National Materials Policy, Office of Technology Assessment, US Congress, August 1974

Erich *Jantsch,* "Technological Planning and Social Futures", Cassell/Associated Business Programmes, London, 1972

M.V. *Jones,* "A Technology Assessment Methodology, Some Basic Propositions", The Mitre Corporation, June 1971

Edward W. *Lawless,* "Technology and Social Shock", Rutgers University Press, New Brunswick, 1978

Joseph P. *Martino,* "Technological Forecasting for Decisionmaking", Elsevier, New York, 1971

R.W. *Peterson,* "OTA Priorities 1979", Office of Technology Assessment, US Congress, 1979

Gordon *Wills,* David Ashton & Bernard Taylor, "Technological Forecasting and Corporate Strategy", Bradford University Press in Association with Crosby Lockwood & Son Ltd., London, 1969

IV Key Issues for Technology Planning and Assessment

1 National Materials Policies versus Market Forces

Previous chapters have the clear message that the materials industry is subjected to increasing pressures and regulations from the socio-political arena. This immediately brings up the question as to what extent a national materials policy, including specific guidelines and regulations, is possible or even desirable. We have to keep in mind that from the supply of basic resources, through the whole materials cycle, to the finished products and their after-use patterns, market forces are the over-riding element and have been successful in creating many self-regulating mechanisms. Therefore, up to now, most leaders of industry and many in government agree that the rule should be: "Keep the free market forces intact wherever possible or restrain them in the form of guidelines and keep governmental interference to a bare minimum."

But, especially in the USA, the question arises whether this is good enough to deal with future challenges. In 1977, the US Federation of Materials Societies sponsored a conference on "Materials and National Issues" where Dr. Frank Huddle, the senior material specialist of the Congressional Research Service, stated the following:

> "Because of the importance of materials for the American economy, we need vigorous technological advance throughout the entire materials cycle.
>
> We need policy goals. We need plans for effective implementation. We need a systems orientation.
>
> What are the elements of this national policy for materials?
> - the rejuvenation of industry through technological innovation;
> - the stimulation of a strong basic research in materials;
> - a closer coupling of the results of basic research with industrial application;
> - a greater flexibility of government regulation to encourage industrial innovation."

Also mentioned were two situations needing correction:

> "An asserted adversarial relationship of government versus industry and something called "crisis mentality" – the idea that we had to wait until things got broken before we took steps to protect them from being

broken, instead of dealing sensibly and rationally with materials problems before they became critical."

Even now, as this manuscript is being finished, the issue of what shape a national materials policy shall have in the USA is still under vigorous discussion and in hearings conducted by the US Congress.

It is not that industry doesn't want governmental guidance, on the contrary, today there is already so much regulation about energy and pollution, which indirectly has a tremendous impact on the materials producing and materials consuming industry like the automotive industry, that this alone is enough reason for government to give the entire materials industry enough guidance so that they can maintain or renew their operations under economically viable circumstances for the foreseeable future. In February 1978, Dr. Julius Harwood, VP Ford Motor Company, wrote the following:

> "The period 1973–1974 indicated dramatically the effects of materials availability and technology upon the industrial health and economic welfare of the USA. Yet, it is not manifest that materials resources and technology have achieved appropriate recognition and visibility in policy and strategy levels of the Government. We see insufficient leadership in administration policy levels with the expertise and perspective of the materials issues and their relationships to other national policies and goals.
>
> We see environmental legislation, regulations, agencies and policies; we see a new Department of Energy and proposed energy programs and policies. But we do not see enough evidence of the recognition, so often articulated in commission and advisory studies, of the intrinsic interdependencies and interrelatedness of environment, energy and resources. Not to comprehend the significance of incorporating these three issues as an integrated resource system for optimum trade-offs stands out as a weakness of our current resources posture. Materials technology is essential to successful achievement of both environmental and energy goals and both of these inturn markedly affect the capacity and capability of our materials producing and using industries."

Dr. Harwood, being responsible for materials innovation in a large automobile company, again stresses the unresolved issue of the problem triangle ("triad") shown on page 5. The same triad which was in the focus of COSMAT and COMRATE recommendations.

By the way, it is interesting to see that in the USA – where more than 80% of the non-fuel raw materials are mined locally – this issue is so dramatically under discussion. In Europe, where on the contrary 80% of the minerals and ores have to be imported, there seems to be rather little concern about the need of material policies and more confidence exists that free market forces and self-regulatory mechanisms will take care of the problem.

In fact, up to now, it seems to be a free market economy "golden rule" that a resource or material shortage which is forecast several years ahead will not happen because prices increase, new supplies will come on stream and furthermore, materials considered critical are used less and less and substitution takes place as was the case with tin.

Dr. Sies of Metallgesellschaft in speeches, in 1978, made this point very clear from the viewpoint of West Germany: "So far the organizations trading ores, minerals and metals have always been able to satisfy the demand because of the self-regulatory mechanism taking place."

Certainly, major disruptive events like a war or a large embargo are a different story, but they would create a specific type of economic crisis where previous government regulations would not help much anyhow.

Therefore, in conclusion, the already existing "over-regulation" in the USA, which cuts deeply into the activities of the materials industry, has initiated a drive for a coherent national materials policy.

2 Energy as a Critical Constraint

As it has been pointed out elsewhere in this book, energy is a serious constraint for the ample, future supply of materials. All basic materials are energy-intensive. In many highly developed countries, the steel industry consumes a sizeable amount of the total primary energy, often between 5 and 10%. Aluminium reduction needs large quantities of electricity, copper and cement need fossil fuels in their production, and plastics utilize feedstock which can be used alternatively for energy production.

In 1976, Goeller and Weinberg published a classic paper on "The Age of Substitutability". This paper explains that only fossil fuels will be scarce in the foreseeable future whereas almost all materials will be available in completely sufficient quantities for an extremely long time, *if* there is sufficient energy. The only exception are a few trace elements and chromium, which is an important ingredient used in making stainless steel. But, there are substitutes for stainless steel. For instance titanium will be an ideal substitute and may therefore have additional importance already around the turn of the century.

Now turning back to energy, the metals industry in the USA consumed around 9% of the total energy in the early 70's and some 90% of this was used for the production of iron, aluminium, copper and titanium.

Of specific importance is the energy-intensity in utilization of leaner ores. Lead is a good example because in several decades, it will have to be extracted from much leaner ores.

As leaner ores are mined to produce one ton of metal, it is necessary to transport, crush and treat larger and larger quantities of minerals and rock. All these processes consume energy and energy-intensive chemicals, and, by the way, could have an enormous environmental impact. Further we must keep in mind that simultaneously, when leaner ores are used, the energy costs will rise as well and this is described in more detail for the case of lead ores, in Figure 19.

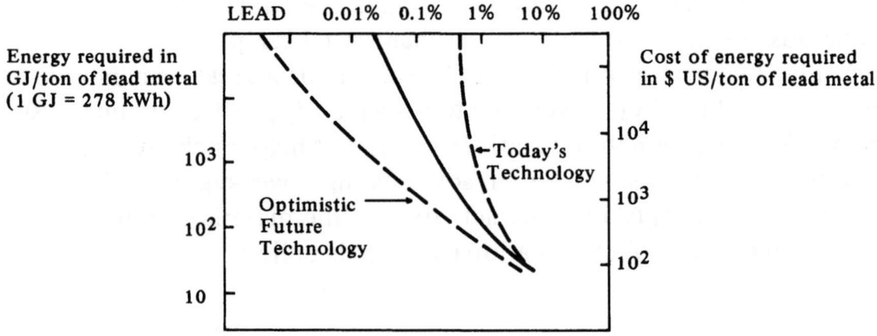

Figure 19: Availability and Energy Cost of Lead versus Ore Grade

Source: Rational Use of Potentially Scarce Metals, NATO Scientific Affairs Division, Brussels, 1976

Bounds on the energy costs are shown and the upper bound corresponds to today's technology, the lower bound to optimistic assumptions about future technologies. Using a mean value in the form of the unbroken line, lead made from common rock could cost up to 1 million dollars per ton. If we make the assumption that a lead price 100 times higher than the 1975 price of $ 200 per ton would be the maximum acceptable price (a car battery would then cost $ 300) the corresponding minimum economic grade of ore is around 0.1%. This would expand the resource base for lead from around 500 x 10^6 tons to about 10^9 tons.

For other metals similar calculations could be made because they occur in dilute form as well as concentrated ores. Copper production in the USA comes into this category, where ore grades below 0.5% copper are now being considered. Chile, on the other hand, still surface mines a 1.2% copper ore. African ores contain 3–5% copper but underground mining is required.

Figure 20 shows the energy costs of copper production with changing ore grades, which indicates that at low ore grades energy cost could become prohibitive.

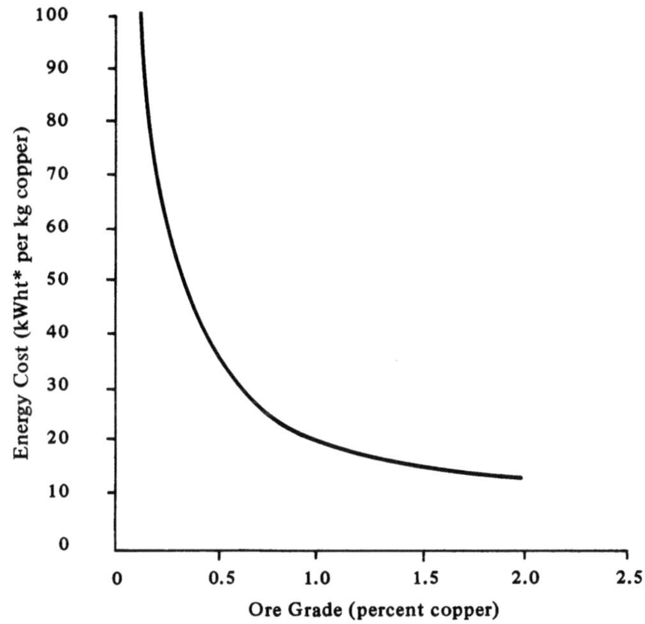

Figure 20: Variation in Energy Cost of Copper with Ore Grade

* kWht = kilowatt-hours thermal

Source: TNO/The Energy Accounting of Materials, Products, Processes and Services, 1976

2.1 Energy Saving in the Basic Metals Industry

Figure 21 shows that the aluminium industry is making great strides to reduce their energy consumption.

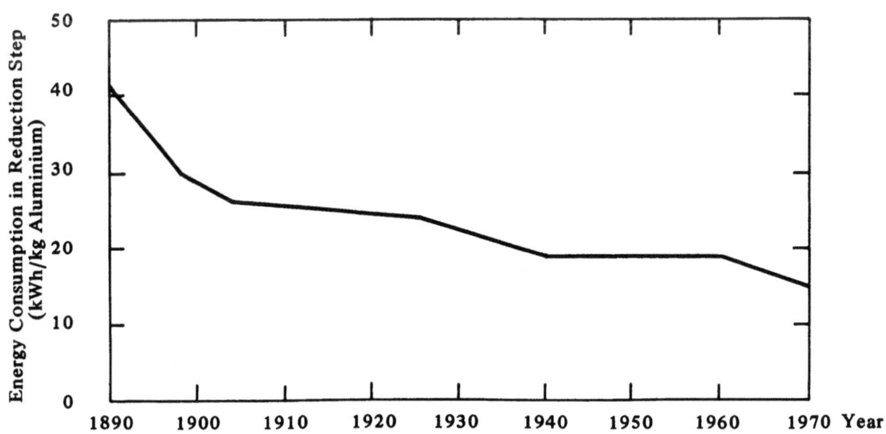

Figure 21: Reduced Energy Consumption by the Aluminium Industry

Source: Ullmanns Enzyklopädie der techn. Chemie, 4. Auflage, 1973

In the future, the aluminium industry is absolutely dedicated to save as much electricity as possible. A new process, developed first by Alcoa, is now under consideration whereby electrolysis of aluminium chloride could reduce energy requirements up to 30% over the currently used fluoride electrolyte. However, until 1990, only a small amount of aluminium will probably be produced via the new process. Elsewhere in this book it was already indicated that plentiful hydroelectricity is a reason to erect new aluminium smelters in the Third World, mainly in Latin America, Southeast Asia and Africa. There, the classical Hall-Héroult process has clear advantages.

Figure 22 shows that, even before the turn of the century, great efforts have been made to reduce energy consumption in pig iron production.

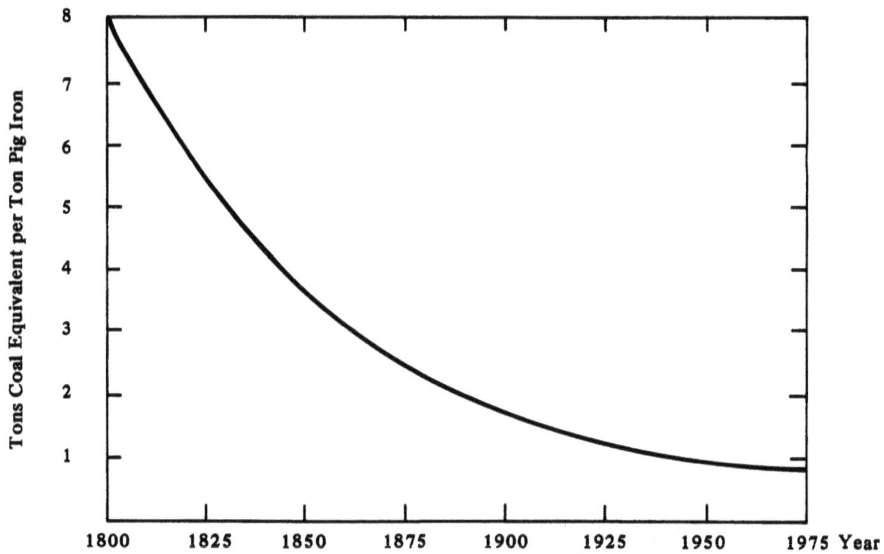

Figure 22: Evolution of Energy Consumption per Ton of Pig Iron
Source: TNO/The Energy Accounting of Materials, Products, Processes and Services, 1976

This was accomplished mostly through economy of scale plus technological advances.

For the steel industry, availability of surplus natural gas and electrical energy is a reason to erect direct reduction units in OPEC countries. Table XXXI clearly shows why: direct reduction uses 3 1/2 times more electricity plus a large amount of natural gas vis-à-vis the classical blast furnaces. Therefore, this scarce type of energy rules out direct reduction in highly developed countries but, not in locations rich in these energies.

Table XXXI: Energy Consumption for Liquid Steelmaking by Several Processes

		BF/ Oxygen Steel	EAF	DR and EAF	$\frac{O_2\,Steel}{EAF} : \frac{80}{20}$
Electricity	kWh	194	460	731	
Natural Gas	Nm^3	–	–	336	
Coke	kg	378	8	–	
Energy for coke making	Gcal/ton steel	0.28	–	–	
Total	Gcal/ton liquid steel	3.10	0.97	4.01	2.67
kWht/ton liquid steel		3600	1130	4660	3110
GJ/ton liquid steel at liquid steel stage		12.98	4.07	16.78	11.19

BF: Blast furnace, EAF: Electric arc furnace, DR: Direct reduction

Source: "Energy Accounting of Steel", Dr. A. Decker, CRM, Liège, Belgium (9th International TNO Conference, Feb. 1976).

2.2 Energy Policy and the Materials Industry

From everything said so far it is obvious that in all countries, where materials industries exist or are to be built in the future, energy availability and energy policy will be key issues. But, such a policy requires as a basis, sufficient accurate information regarding future energy requirements of the materials industry. For long-term planning, there is a need to correlate energy policy in a systems approach with many segments, like:

– available primary energy sources for at least the next 20 years;
– market-pull for the produced materials, possibilities for exports;
– cost-benefit-risk analysis;

– technology assessment, because the energy and basic materials industries have strong interaction with ecology.

Energy analysis is a rather new discipline. The first detailed results were published around 1972/73. The energy analysis of energy production itself became an important issue. Experts disagreed often on how to do energy accounting because energy is not always equal energy: to quote just one example: hydroelectricity in an isolated part of the world is not the same as electricity in an utility system where, during peak load times, blackouts occur. Today, input/output tabulations for all basic industries exist in most of the developed countries, and are regularly brought up-to-date. Behind this is a complex process analysis to bring all energy inputs into comparable terms. Although great efforts have been made no energy accounting method seems to be always correct. Energy analysis of the materials industry may be more promising in one sector than in another. For instance, there may be confusing results comparing two different materials like steel and aluminium just on the basis of energy input during processing. Properties of different materials during their whole life-time may be the overriding factor, even from the viewpoint of energy conservation.

In the next section of this chapter, energy accounting is described.

3 Energy Accounting of Materials

By Dr. D. Spreng, Swiss Aluminium Ltd

"Energy is the ultimate raw material"*

The economic and environmental cost of energy is likely to increase in the future. Wise energy use is therefore a guiding principle for all promising developments. The first step to wise use is proper energy accounting. The present discussion deals with some issues in energy accounting within the materials industries.

Although energy itself is an objectively measurable physical quantity, energy accounting can be done in many different ways. In the recent past, energy accounting often looked more like a game than a science.

For every kind of activity and for every product, the corresponding amount of expended energy can be calculated. If one travels by car, a passenger-kilometer requires approximately 4 Megajoules (MJ). Travelling by train one might get 5 times, or by bicycle 30 times, further with

* A.M. Weinberg

the same energy expenditure. The 4 MJ could also cook a meal, bake a cake, or keep your room warm for one hour. The game now consists of calculating the numbers, adding them up and drawing conclusions. The result obtained depends on the method used.

Take for instance the above statement that it might take 5 times more energy to travel a certain distance by car than by train. The calculation can be based on any assumed numbers for train occupancy, vehicle power and speed. Then you can include or omit energy expenditure for rail and road building, train and car construction, maintenance, and auxiliary installations. When discussing your result you can stress the relative merits of the two modes of transportation any way you like. Thus, because energy accounting is not clear cut, it may not only be used to gain insight but also to provide ammunition in support of preconceived ideas.

3.1 Aluminium in the Energy Accounting Game

Aluminium has often been a subject of the energy accounting game. In the early 70's an interesting controversy developed between two prominent American ecologists. It started with Barry Commoner and his book "The Closing Circle". He contends that, for the US in the last 30 years, productive technologies, having intense impacts on the environment, have displaced less destructive ones, and that it was this counter-ecological growth pattern which caused the environmental crisis. For instance, he says, new technologies with high growth rates are in general more energy intensive than older, established technologies. The high growth rate of aluminium is given as a prime example of a "technological flaw", since "aluminium requires for its production about 15 times more fuel energy than steel".

Paul Ehrlich, author of "The Population Bomb", together with John Holdren wrote a critical review of "The Closing Circle". They maintain that it was not faulty technology which disturbed the ecological balance but simply the amount of technological activity of a growing population: "Many of the technological flaws we all deplore will not easily be corrected, for the alternatives too, are full of defects". On aluminium they write:

> "It is a popular misconception that most aluminium goes into beer cans. Actually, containers of all kinds account for 10 percent of aluminium consumption, while the building and construction industry uses 23 percent. In the building industry, aluminium is replacing wood for siding, window frames, awnings and other applications, not simply because it is cheap, but because it is durable and maintenance free. Many people believe low maintenance is part of affluence. Aluminium is costly in energy, but meeting the demands of a growing population for better housing with wood alone would put an awesome demand on forests already being too intensely exploited. (Need we remind Dr.

Commoner that "there is no such thing as a free lunch?") In transportation, aluminium replaces iron in automobile engines and steel in aircraft. Aluminium is about five times as costly in energy as steel to produce, but lighter cars and aircraft burn less fuel. In the electrical industry, abundant aluminium is replacing scarce copper as a conductor of electricity. This is not coincidence or technological frivolity – it is the classic example of the sort of substitution that is inevitable when a growing, affluent population presses on a finite resource base."

3.2 Some Basic Rules for Approaching Energy Accounting

Today the energy accounting game has developed into a discipline. The most commonly employed method is called process analysis. In process analysis, first the process and its limits are defined and then all energy expenditure (within the defined limits) is taken into account. But there are three main problem areas: firstly, to set the boundaries in accord with the aim of the study, secondly, to add up various types of energy and thirdly, to deal with branches in the process-chain. These three areas contain rather difficult problems which readily explain most of the discrepancies in energy accounting.

One can find any number from 45 MJ to 300 MJ as the energy requirement to produce one kilogram of aluminium, depending on whether the energy figure just refers to the electrolytic cell, or whether it includes the bauxite mine, the alumina plant, losses at the oil fields, transportion, the anode plant, the caustic plant, the smelter (including perhaps a coal mine and a coal-fired power station of low efficiency), and materials used for plant construction.

In addition, it should be stated how various types of energy are added up and how branches in the energy chain are dealt with; an example of a branch being electrolysis of sodium chloride (electricity coming in, caustic soda as well as chlorine and hydrogen going out).

The production of aluminium requires thermal energy (mainly for calcining alumina) as well as electrical energy (mainly for electrolysis). How can these types of energy be compared and combined to provide an absolute number suitable for a total energy calculation? One possibility would be the use of direct physical conversion factors. It would be appropriate for hydropower to say 1 MJ of electricity equals 1 MJ of primary energy.

When electricity is produced in thermal power stations, it would probably be better to use a conversion rate of 3, i.e., three units of primary energy are equivalent to one unit of electrical energy. However since, worldwide, the aluminium industry is based on about 50% hydropower, for general considerations a conversion factor of 2 would probably be more appropriate (the average between hydro- and thermal power electricity generation).

Process analysis can provide meaningful energy requirement comparisons for one product as compared to another equally useful product. Besides defining all processes in a sensible way, using well-defined energy conversion factors and taking care of branches in the process-chain, the following factors have to be considered: product-life, energy-requirement while using the product, and energy expended for disposal, reuse, or recycling.

Another method of studying energy expenditure is borrowed from economics. It is based on input-output theory, for which Wasily Leontief received the Nobel Prize. Here the economy is divided into sectors and the influence of one sector on another is calculated from the inputs and outputs of each sector. A very extensive data bank and large computer models are required. Nevertheless, the same rules hold for this methodology as for process analysis.

In summary, energy accounting should be done according to the following three golden rules.

1) *Compute total input of primary energy* from the mine to the finished product with appropriate boundaries and conversion factors.
2) *Take into account the energy saved by recycling.* In case a material can be recycled, a large part of the energy required to produce the primary material may be recovered and thus the material represents an "energy bank".
3) *Use energy accounting for comparing real alternatives* (products or services) and include energy saving or energy use during the products entire life cycle.

Today energy accounting is considered a serious discipline and a dominant factor in determining the environmental impact of a product and is taken seriously by lawmaking bodies. Furthermore, energy accounting is a useful tool for corporate strategy.

The following section will deal with some applications in transportation where both steel and aluminium can be employed as real alternatives.

3.3 Comparison of Aluminium and Steel Applications in Transportation

Following the first "golden rule", we compute the total input of primary energy from the mine to the finished product. As boundaries we take the property boundaries of the various factories and we account for the energy (fuels and electricity) which crosses these property boundaries (e.g., we include not only energy used for the various processes but also for the infrastructure of the factories), we include also energy for necessary transportation and the energy requirements of important materials used for production. We do not include here energy for constructing the factories.

In Figure 23 the total input of primary energy from the mine to the finished product (here 1 ton of sheet metal) is given for aluminium.

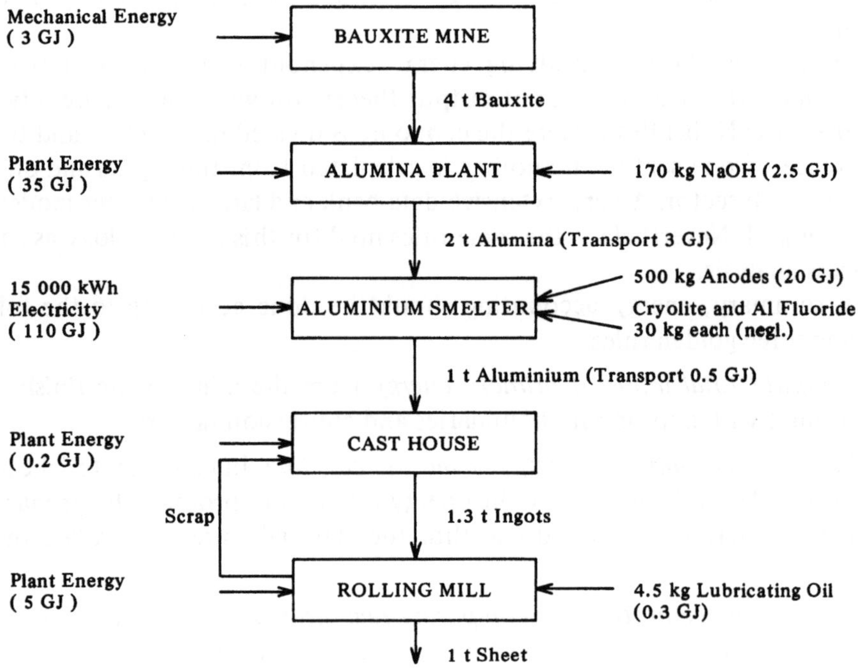

Figure 23: Energy Requirements to Produce 1 Ton of Aluminium Sheet

The energy requirement for steel is given in tabular form (Table XXXII). However, a direct comparison between the energy required to produce 1 ton of aluminium sheet (179.5 GJ) or 1 ton of steel sheet (31 GJ) should not be made unless the three golden rules are adhered to.

In the case of steel production, coal has been the preferred source of primary energy but oil and electricity from any suitable source are used more and more. The aluminium industry now seeks locations for its smelters with electric power which is not costly both economically and environmentally. Usually it will be hydropower in relatively remote areas (in so-called "electric islands").

We now come to the second of the golden rules and consider the "ener-

Table XXXII: Energy Requirements to Produce 1 Ton of Steel Sheet

	Electricity kWh	Primary Energy GJ
Ore (57 % Fe) Mine		0.2
Coal Mine		0.7
Ore Transport		1.4
Pelletizing		3.3
Blast Furnace		16.2
	70	0.7
1 ton of iron		22.5
Steelmaking (including dust filters, oxygen making and slab casting)	100	1.1
1 ton of steel		23.6
Rolling Mill		4.4
Misc. Energy		3.0
1 ton cold rolled steel sheet		31.0

Sources: Stahl-Eisenkalender 1970–74, Lurgi-Handbuch, VIK-Jahresbericht 1972, Jahrbuch für Bergbau und Energie 1973.

gy bank". We will do this by quoting from an address by Dr. Paul Müller*, President of Swiss Aluminium Ltd.:

"Today, in Europe, the tonnage of aluminium recycled is equal to more than 28% of the tonnage of primary aluminium produced, this value is somewhat lower in the USA. Accordingly, efforts are being made to promote recycling in the USA; by way of example, it is planned to return more than half the beverage cans to metal circulation. If this succeeds, it will mean an increase in secondary aluminium of about 500 000 tons per year.

Depending on the application, the life of an aluminium product can vary considerably. While beverage cans may have a life of 6 months, aluminium used in building and construction, transportation and engineering can have a life of several decades. The average life of aluminium products may be assumed to be ten to twelve years (Table XXXIII).

* "Aluminium and Energy", Share Holder's Meeting, Zurich, April 19, 1978.

Table XXXIII: Life Span of Aluminium Products

Aluminium Application Sectors	Market Share in %	Product Life in Years	Recovery Rate in %
Transportation	20–30	10	30–50
Building and Construction	11–26	10–30	40–70
Packaging	10–15	1	3– 5
Electrical Engineering	7–16	10–30	60–70
Consumer Durables	6–14	4–12	5–20
Machinery and Equipment	6–10	10	30–50

The expected decline in the growth rate of primary aluminium production on the one hand and – particularly in the USA – the additional recycling of aluminium from vehicles and cans on the other, will increase the importance of secondary aluminium. With optimal development, the tonnage of secondary metal may reach about 35% of the primary aluminium production in the Western World within the next 10 to 20 years, as compared to 28% today."

Now, let us proceed to the third golden rule. We quote again from the same speech:

"Outstanding examples for energy saving, with aluminium products, can be found in all four major application sectors of aluminium, that is, in:

- electrical engineering;
- building and construction;
- packaging;
- transportation.

But the most impressive and convincing demonstration of the energy saving effect of aluminium is its application in the transportation sector. Here, the functional properties of aluminium may be used to their fullest extent and it is easy to show energy savings in numbers. In the design of passenger cars, European, Japanese and US manufacturers are examining steel components, part by part, to determine the feasibility of using aluminium with the objective of reducing the weight of their vehicles. Weight reduction results in noteworthy fuel savings and a positive energy balance (Table XXXIV).

Table XXXIV: Energy Saving Possibilities – Passenger Cars: Due to Weight Reduction by Using Aluminium

Aluminium Used per Car		Gasoline Saved per Car During Its Life
Today in the USA	50 kg	850 liters
In the Future	100 kg	1700 liters
	150 kg	2600 liters

The use of aluminium per vehicle is increasing steadily. The applications include rolled and extruded parts for bumpers and the body. It should be noted that the use of aluminium for castings dominates at present. Since castings use substantial amounts of secondary metal, this increases the energy saving effect.

In commercial vehicles, aluminium reduces weight and increases payload. The increased payload and lower weight pay for the higher price of an aluminium vehicle after just a short time of operation.

For example, in rail transport, a hopper car developed to carry phosphates and other bulk materials, saves 5 times the extra energy that was initially required for the aluminium used in its construction, rather than steel, over a life span of 20 years.

Suburban and, more particularly, subway and rapid-transit trains are outstanding examples of the functional and appropriate use of aluminium. In these applications the energy savings, due to reduced weight, assume particular significance because of frequent, rapid acceleration to high speeds.

Recently, the subways of Vienna, Brussels and Paris were equipped with cars with a high aluminium content. This trend will continue."

3.4 Why Could Plastics Survive the Quadrupling of Oil Prices?

Plastics entered their boom years in an era of cheap energy. The quadrupling of oil prices 1973–74 triggered fears that plastics, using oil and/or gas as feed stock, would increase in price so much that they could not compete successfully with substitute commodities anymore. In 1974, energy accounting suddenly became of great interest to each plastics producer.

As it turned out energy accounting for plastics is particularly difficult. The difficulties stem from the fact that the production of plastics cannot be described as a simple chain of processes, but must be seen as an interlocked network with many junctions and branches. This is particularly true in the case of plastics which are produced from oil rather than from natural gas. The cracking of naphtha is tied to the previous process step, crude oil refining, as well as to other processes by many energy carrying connections (pipes for crude oil, heavy oil, intermediate products, steam as well as electricity conductors). One possibility is to charge all products of refining and cracking an equivalent amount of the feedstock and energy requirements from the over-all process. For instance one can use the equation: energy requirement for olefins equals their heating value plus 20%. This kind of definition implies, that all products are equally desirable, which is of course hardly the case. However no other way of doing the calculation is any more satisfying. For polyethylene, (both high and low density), the heating value of ethylene plus 20%, plus the direct energy requirement for the polymerization results in a total energy requirement of 85 MJ/kg. In comparison with polyethylene the energy requirements for PVC are somewhat lower, for polypropylene a little higher and for polystyrene quite a bit higher.

Comparing the energy requirements of typical plastic products with similar products made from some other material yields a surprising result. In spite of the fact the plastic products consume energy (e.g., oil) both as raw material and as energy for their production, their total energy requirement is often of the same order of magnitude as their substitute products. In Table XXXV the energy requirements for 3 types of sewerage pipes and 3 types of water pipes are listed. We can see that the energy requirements for plastic products are quite similar to energy requirements for comparative materials. It is interesting to note that for the production of the alternative materials also fossil fuels are required, i.e., it takes about the same amount of oil to make a steel-reinforced concrete pipe as an equivalent PVC pipe.

Table XXXV: Energy Requirements for Pipes

Sewerage pipes (40 cm dia)	GJ/meter
Reinforced concrete	0.9
Clay	0.7
Polyvinylchloride (PVC)	1.0
Water pipes (30 cm dia)	
Polyvinylchloride (PVC)	1.3
Polyethylene (PE)	1.3
Cast iron	1.1

A special situation is the replacement of plastics by wood for building applications. What energy requirement should be assigned to wood, a very small one for sawing and transporting or a larger one which includes the heating value of wood? Energy accounting does not yield any conclusive results here.

In summary, plastics could survive the quadrupling of oil prices for three reasons.

- Many alternative materials require almost as high an energy (often oil) input as do plastic products.
- Plastics are either produced from natural gas or from refinery products, which are jointly produced with gasoline and oil for heating purposes etc. To be profitable, the refinery needs a certain income from all products put together. The income from particular products can be adjusted according to the market situation.
- Before 1973, some plastic products were so much cheaper to produce, to handle, to install etc., than the products they were replacing, that a certain price increase did not drastically change the competitive situation.

However, for future considerations, it must be kept in mind that plastics are made from the light fractions of crude oil cracking – for which there is great competition. Therefore, the plastics industry must plan for continued drastic price increases of their raw materials.

3.5 Some Energy Issues in Packaging

3.5.1 One-way versus Returnable Beverage Containers

The beverage container example is the classical energy accounting problem. One reason for this is that "throw-away" beverage containers create an obvious waste problem and since energy as such is not visible, the visible waste heaps of beverage containers represent an image of wasted energy and raw material. The second reason for the prominence of this classical energy accounting problem is that it is a rather rare case of a product where several materials practically do the same job and where there are no particular preferences for one or the other material. In the preceding example of the water pipes for instance, plastic pipes have several advantages (flexibility, length, lightweight) which are difficult to express in terms of energy, but which weaken the energy argument.

Table XXXVI shows one such beverage container calculation.

Table XXXVI: Energy Expenditure for Beverage Containers (carbonated beverages and beer in Switzerland)

	MJ/l beverage
Glass bottle, large size (returnable*)	0.8
Glass bottle, normal size (returnable*)	1.5
Glass bottle, normal size (one-way)	10
Tin can	7.5
Aluminium can	12
with a recycling rate of 50 %	8.5
PVC bottle	6.5

* With 20 trips

Source: Schweizer Aluminium Rundschau, No. 4, 1975

What can we learn from that result? The result is simply that in a country with short transportation distances, well organized return-transportation (on otherwise empty trucks) causing no additional energy expenditure, for the case of carbonated beverages and beer, returnable glass bottles with a high density of use are rightly being used.

The energy expenditure can be entirely different when any of the following input data is changed:

- type of beverage;
- frequency of consumption;
- density of consumers;
- transportation distances.

Not to speak of other considerations such as sporting events where broken glass could be dangerous and hikes, where the extra weight of heavy glass bottles is not very comfortable etc. Other calculations, i.e., for non-carbonated mineral water in France, where today mostly large, thin-wall, one-way plastic bottles are used, show a large energy saving compared to returnable glass bottles.

Energy will not be a severe constraint on the development of beverage containers, since the energy requirement for competing systems are rather similar. It is hoped that "the beverage container calculation" is not used out of context for setting up legislation, which might impair the development of the best system.

3.5.2 Recycling – Does It Save Energy?

In assessing the value of recycling, it becomes particularly obvious how incomplete an economic analysis based on today's prices is. Raw material prices are fixed at an arbitrary level. Of course often, the level is determined by supply and demand, but should today's supply not be logically related to the supply available to future generations? The cost of raw materials does not systematically include the cost of resource depletion to society. In the same way, the cost of waste disposal is not included in the price of products. This, of course, should be the case, if a consumer is to make a positive contribution to the economy when he purchases something. It is clear that the high social cost of resource depletion and waste disposal inspires a great interest in recycling. However today's market system may not encourage recycling sufficiently. There is not only the basic problem of accounting for social costs, but, in addition, the many tariffs, freight rates and regulations which discourage rather than encourage recycling. This is why recycling is often assessed in terms of energy accounting.

The remelting of metals requires much less energy than the production from virgin ores or from natural metal compounds. Figure 24 illustrates this for the case of aluminium.

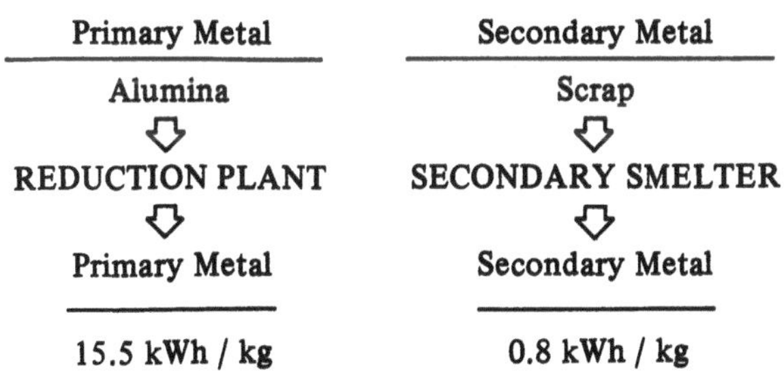

Figure 24: Consumption of Electric Energy for Primary and Secondary Production of Aluminium

Remelting aluminium scrap requires about 5% of the energy required for the reduction of aluminium oxide. From this, it is often concluded that recycling aluminium saves 95% of the energy that is required to produce it from virgin ores. Although this is often the case, it is not necessarily so. The energy expenditure of a remelting facility is not a constant figure, but depends on the quality of the incoming scrap. Very thin gage scrap will have more burn-up, and perfectly clean scrap can be melted in an electric furnace whereas other scrap is processed in a salt furnace. The reprocessing of the salt, if it is done, requires a substantial amount of energy, the amount of which increases with increasing impurity of the charge.

But, recycling is more than remelting, to point out the difference, let us consider an extreme case. If an aluminium can is recycled by bringing a single can to the nearest recycling center, making a detour of 1/4 mile by car, this results in an energy expenditure, through the use of gasoline, which is as large as any possible energy saving from recycling. Although few people might be so ardent a recycling fan as to drive 1/4 mile with one can, recycling involves many steps and each step has to be managed efficiently.

It is a matter of simple arithmetic to determine that an aluminium-plastic laminate with less than 30% aluminium should preferably be burned in an efficient incinerator-energy-recovery-system, to recover the maximum energy possible, rather than be processed in a scrap cleaning furnace at the secondary aluminium plant. The incineration of plastic films or laminates to recover the heat content has a drastic influence on the energy picture plastic presents, for packaging applications. The packaging application can be viewed as a *precycle* to the end use of crude oil in the production of heat.

One ton of plastics requires roughly two tons of crude oil. Precycling one ton of plastics therefore requires at least energy equivalent to about one ton of crude oil. It should however be born in mind that plastics in the waste stream are not as readily usable as oil in a barrel and very often solid waste has to be disposed off whether we need heat right then and there or not.

The more energy will become a constraint to our endeavors, the more recycling of metals and precycling of plastics will increase. But, development of sensible systems takes time.

3.5.3 How Packaging Saves Energy

The function of packaging is to assure the delivery of a fixed amount of goods, unspoiled and unharmed from one point to another and/or to preserve the goods for a desired length of time. Although used packaging contributes typically about a third to household waste, the prevention of waste, particularly food waste, is a most important function of packaging. Protection from spoiling, in particular, can often be expressed as a saving in energy and this is often many times greater than the total energy expenditure needed for the production of the package.

For example, the protective action of aluminium foil for butter, expressed in energy units, is many times greater than the energy expenditure for the foil production. Countless other examples show that well-designed packaging helps to shape our food system efficiently from the point of view of energy expenditure. Compared with earlier days, efficiency has essentially increased: foodstuffs nowadays are more capable of transport and storage. Therefore spoilage has been greatly reduced.

One possibility for quantifying the "useful value" is shown in Figure 25.

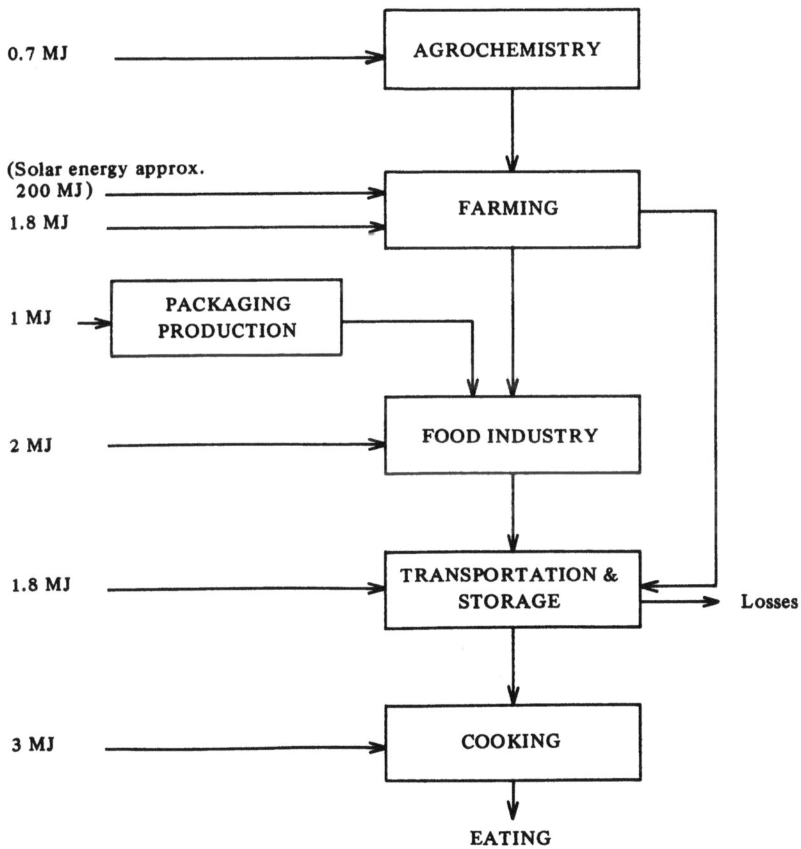

Figure 25: Daily Per Capita Energy Expenditure for Food in the USA
Source: D.T. Spreng, Kunststoffverpackungen und Umwelt, ASKI, Zurich, 1975

Without the expenditure of 1 MJ, which is the daily per capita energy needed in the USA for total packaging production, a good proportion of the 6.3 MJ used for technical operations in agrochemistry, farming, the food industries, transportation and storage, as well as some part of the sun's energy stored in the agricultural products (total about 200 MJ) would be lost by product deterioration.

Of the solar energy used in photosynthesis, "only" a few percent are converted into carbohydrates and, in the subsequent processing stages up to the final food product, part of this energy is lost. The dietary energy value of a foodstuff, therefore, cannot be read off directly from Figure 25 and is not directly comparable with the technical energy expenditure. It is a highly "refined" type of energy, which is associated with the carbohydrates and proteins contained in food products. The use of crude technical energy in the manufacture of packaging for the protection of this very valuable energy is therefore of great significance.

The reference term "benefit" in a cost-benefit-analysis should always relate to something directly of value or agreeable to the consumer, thus, not to packages but to freshness of foods, not to a certain number of joules but to a direct benefit. If very different systems are compared with each other, care must be taken that the "benefit" is chosen precisely enough for the user concerned. The purpose of a food system is to produce the maximum benefit to the consumer at minimum costs (including energy expenditure, raw material expenditure, cost caused by litter and pollution of all kinds). The package is thus only a part of the total system to be optimized.

3.6 Materials in Energy Supply Units

There is one specific important crosslink between materials and energy: the.energy accounting of new energy supply units. Debates have been carried out on the subject in the last 5 years or so and many divergent opinions were offered on how much energy for instance has to be advanced to build a nuclear power plant before net energy production starts.

We have to realize that the conventional energies like oil, gas and electricity will considerably increase in their price world-wide. Then, the question arises, how will this effect the priority, costwise, of alternative energies. It turns out for instance that a unit to extract oil from shale or tar sands, or for coal gasification, needs an enormous amount of materials and their price and energy requirements are a dominant factor determining the cost at which the alternative energy can be provided. When oil had a price around $ 3 a barrel we often heard that oil from tar sands could be made at $ 8 a barrel. When oil rose to $ 12 a barrel we were told that the oil from shale would cost $ 18 a barrel and so on. It may be quite possible that the cost of construction material will raise in line with cost of oil and therefore alternative energies may not be competitive to oil, in the foreseeable future, for this very reason.

Critics of nuclear power have claimed that some plans for developing nuclear power were examples of fast growing systems which over a few decades would require more energy than they supplied. Critics of solar energy are claiming the same for solar energy programs.

Of course, any energy producing system should produce much more available energy than it consumes. This certainly has to be the case in order for an energy producing system to be economically viable. Such a system has to pay the cost of all the inputs from the sale of the energy output. The cost of the energy input is only one of the inputs (very often amounting to about 10%).

Taking as a given fact that the energy producing system does produce more available energy than it consumes, the arguments of dynamic analysis have therefore limited scope. Any endeavor requires an initial input: One has to seed before one can harvest, one has to invest before the return on the investment makes the investment worthwhile.

In practice one will find that energy producing systems that make technical and economic sense will be net energy producers and that programs which can actually be realized will after several years certainly produce net energy.

4 Substitution and Conservation of Materials

Since the early 70's, a world-wide drive for conservation of materials and energy has been in the focus of public, government and industry as well.

Today, in the highly developed countries, there is less materials use per unit of output. One reason for this was shown in Figure 5 on page 15 – the rapid growth of services, the "tertiary sector". But there are other reasons for less primary materials input per unit GNP in the future. Increased reclamation of materials from waste or scrap will emerge. But it will take years before there is an additional significant contribution to total materials input because of given redundancies and due to free market forces favoring materials of primary origin.

In Figure 26, a brief description is presented how engineering materials can be conserved either by substitution or less dissipative use. We will deal with both groups of measures.

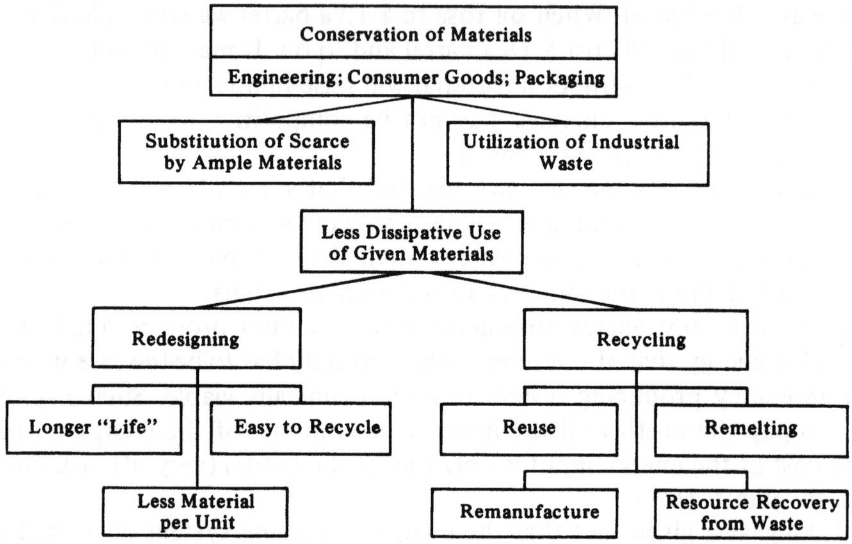

Figure 26: Conservation of Materials

4.1 Selection and Substitution of Materials

Substitution goes back over millenia, the Stone Age was followed by the Bronze Age, which in turn was replaced by the Iron Age. Today we are entering the age of man-made and/or reused materials. Substitution is ongoing and very intensive between metals and non-metallic materials. The main incentive for substitution may be purely economic or better functional properties of the "winner" in this race. The resulting advantages may be obvious for the manufacturer and/or the final user.

We will limit ourselves to just a few case histories, to explain how quick and surprising materials substitution takes place nowadays and what the decision criteria are. Afterwards, we will try to point out a few basic rules which apply to substitution in general.

A. Building and Construction

Because of the need to have better insulated houses in many highly developed countries and because of the trend to more prefabricated systems drastic changes are taking place in the choice of materials either for new buildings or for retrofit of existing buildings. For instance, in West Germany the market for plastic window frames shows a rapid expansion which is mostly due to retrofit of existing buildings but partly due to the improved properties of the plastic frames (Figure 27).

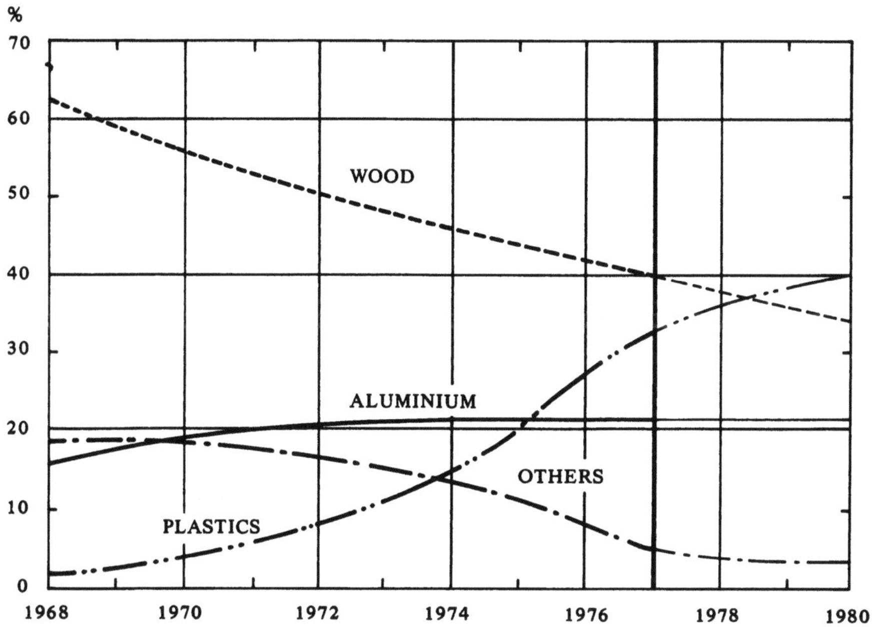

Figure 27: Percent Market Share of Window Frame Materials in West Germany
Source: Chemische Werke Hüls.

Concrete products are making rapid improvements in the form of foamed concrete with good insulating properties or by thin-wall reinforced concrete sections, which can compete well with steel profiles.

On the other hand, sheet steel has improved drastically in corrosion resistance through better galvanizing and/or organic coatings and therefore now competes successfully against other materials for roofing and siding of homes or industrials buildings. Energy accounting is a main element in the decision process for materials preference. This means, in a cost-benefit analysis, such materials will be preferred which result in energy savings or such components where solar energy can be incorporated into house design to bring solar heat into action, either by vertical, horizontal or inclined surfaces which are an integral part of the house.

B. Other Sectors

Packaging of beverages and food is a very competitive market. This favors substitution on purely economic grounds. Furthermore, there is increased legislative pressure to penalize throw-away packaging. This initiates another group of substitution processes which, in the USA for example, favors aluminium can recycling.

Another industry with intensive substitution is the automotive industry. In the past cost was the main reason, now, because of government directed fuel economy standards, the US automobile producers are evaluating components on a weight basis as well as from a cost standpoint.

In both the packaging and automotive industries, energy consumption during the entire lifecycle of the product is now a main driving force for substitution.

We could go on quoting cases of substitution. Instead, we will now briefly go into some fundamentals.

4.2 Substitution in the Focus of Technology Planning

In our book, it is shown repeatedly that in the future, the total environment of the main materials industries will be a decisive force for the amount of primary materials to be used and for the different approaches to materials conservation and reuse.

Therefore, the methodology of technology planning, as described in Chapter III is a useful tool to develop strategies and resulting action to deal with substitution processes not only by reacting, but by better early recognition of threats and opportunities.

The conditions for forecasting substitution processes can be briefly summarized in the following three points.

A. Forecasting of technologies and products.
B. Forecasting of future needs and acceptance.
C. Understanding of the constraints and opportunities in the total environment of the materials producing and using segments of the society (socio-political constraints, like energy, environment, siting of new industries).

We can assume that it is rather simple, or at least possible with sufficient accuracy, to predict when a certain technology could be ready for use (A).

But the second criterion (B), the forecasting of future needs and resulting demand is somewhat more difficult, and the third, (C) forecasting and understanding the future environment of the industry, is extremely difficult.

Therefore, a technological forecast (A), that a certain technology (example: fast breeders) could be ready ten or twenty years from now is not of much use, if the decisive elements (B) and (C) are not there, that means the acceptance by the public or other socio-political conditions.

However, if a substitution process has neither impacts in the socio-political arena, nor enters the decision-making process of the consumer, other than by price, performance and quality considerations, substitution of materials may take place rather rapidly. But, as soon as the before mentioned elements (B) and (C) enter the picture, substitution forecasting becomes more complex and also the changes may occur over a much longer time interval. This is explained in the following subchapter which describes the substitution ladder.

4.3 Seven Levels of Substitution

Table XXXVII describes the structuring of substitution as proposed by R.U. Ayres. This "substitution ladder" deals with the questions: "Under what circumstances does technological substitution occur? and, What are

Table XXXVII: The Substitution Ladder

Level (Rung)	Brief Description	Examples
VII	Shift in social or personal values or goals resulting in shift in demand	More consumer goods versus quality of life
VI	Shift in strategy to achieve goals	Telecommunication versus personal travel
V	Shift in technical means (i.e., systems) to implement strategy	Individual personal transport versus mass transport
IV	Shift of subsystems, within a system (design change)	Internal combustion engine versus battery powered vehicle
III	Shift in components (design change)	Piston engine versus turbine engine
II	Shift in materials for specified component	Aluminium versus cast iron for engine blocks
I	Shift in materials processing technology	Ingot casting versus continuous strip casting of metals

Source: Robert U. Ayres (partly modified)

the basic socio-economic mechanisms that enhance or discourage substitution at various levels? Table XXXVII shows, that substitution may happen in seven different levels. Level number I explains substitution due to advantages by going to a different process in a basic materials industry.

Our book has numerous examples for level I substitutions. They are now often initiated by energy constraints like Hall-Héroult electrolyte versus aluminium chloride in aluminium smelting or blast furnace versus direct reduction in iron making where smaller plant sites and availability of surplus natural gas are incentives for choosing the direct reduction process.

If we go up the ladder, demand forces gain importance; in the lower part of the ladder, supply forces are dominant. A key element is that substitution at any of the higher levels in most cases can trigger substitution at all the lower levels. For instance the obviously existing trends for substitution in levels V, VI and VII will result in less materials use in level I, the materials processing industry. Other cases which prove this are the examples quoted in levels III and IV. For instance, conversion to a new car engine design would have an enormous impact on suppliers of production equipment, to stockholders, the whole maintenance system and even the petroleum industry. Another observation is that the frequency and speed of substitution is much greater in levels I–III than in the higher levels, where changes would involve far greater socio-economic inertia. Therefore, substitutions on levels V–VII need a correspondingly greater driving force, whether it may be from the side of energy constraint, government regulation or market pull. In any case, it can be concluded that changes at the highest levels of the "ladder" will be always slow. The motivation for substitution in levels I–VI are either economic or functional advantages. On levels IV and V macro-economic or political reasoning could be the driving force. On levels V–VII an increasing number of psychological elements are to be recognized. Mass media play an important role here. The rate of change in our societies may be accelerated as pointed out in the book "Future Shock", by Alvin Toffler.

There is another interesting observation by Ayres: the ease to carry out substitution at the lower part of the ladder reflects the tendency of our modern civilization to solve its problems with the smallest amount of disturbance. This is not only true for society as a whole, but even for the decision making process in the materials industry. Many decisions in industry are made for a time horizon of 3 to 8 years. Thus maximizing on short-term advantages at the expense of long range possibilities.

4.4 Introduction to Materials Recycling

By Professor Michael Bever, Massachusetts Institute of Technology

The recycling of materials has been considered in recent literature from the technical, economic, ecological and institutional viewpoints.

Recycling is an integral component of the production-consumption system. Materials production, manufacturing operations and the discarding of obsolete goods present opportunities for recycling. By its very nature, recycling adds to materials supplies and tends to alleviate resource depletion. In so doing, it eliminates residuals of production and consumption which would have to be disposed of in other ways. If carried out wisely, recycling can also bring about process savings, especially in regard to equipment and energy requirements and pollution abatement.

4.4.1 Recycling in the Production-Consumption System

Recycling takes place at three stages of the production-consumption system of typical materials, as shown by the loops in the flow sheet (Figure 28):

1) the production stage
2) the manufacturing stage, and
3) the post-user stage.

These three stages are characteristic of most metals, and of most of the packaging materials.

The three recycling stages play distinct and different roles in the flow of materials. They interact with each other in ways which exhibit typical systems behavior and form the "recycling system", which is a subsystem of the material system.

The material recycled in each stage is designated by a specific term, in the case of metals: (1) home (revert or runaround) scrap, (2) new (prompt industrial or processing) scrap and (3) old (post-user or obsolete) scrap. In some of the following discussions the term "scrap" will not be restricted to metals, but will apply to secondary materials generally.

The three classes of scrap can be ranked by criteria such as quality, time lag before reprocessing, and extent of recycling (or "recycling intensity") as shown in Figure 29. This ranking is reflected in the competitive position of the three classes of scrap.

Home scrap has the highest priority for use and is returned directly to the production process. In general it is the most desirable kind of scrap, since it has a known composition and is usually uncontaminated. The preferential position of home scrap affects the demand for new and old

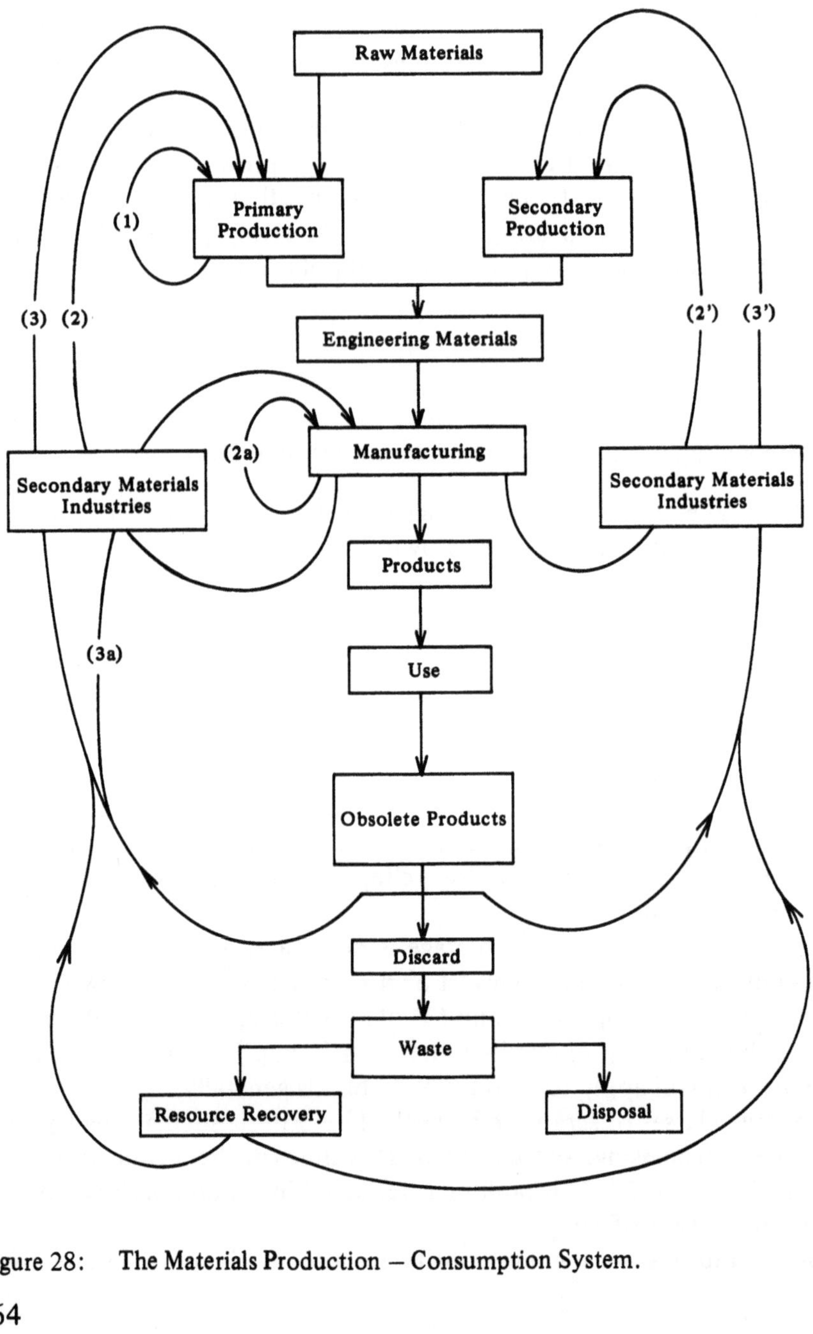

Figure 28: The Materials Production – Consumption System.

("purchased") scrap. This is especially true in the case of primary production processes that have a limited capacity for utilizing scrap such as the basic oxygen process of steelmaking (BOP).

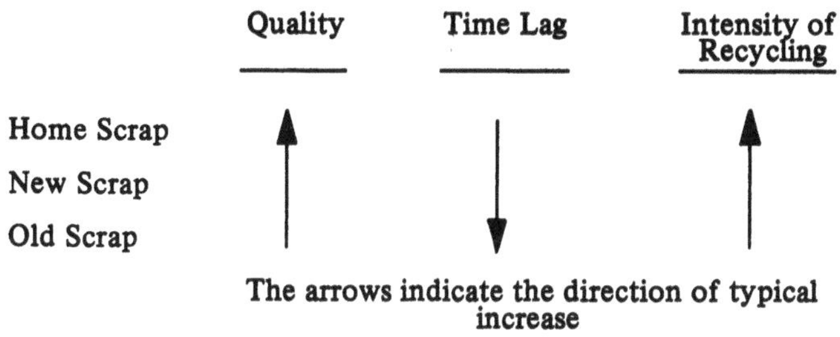

Figure 29: Typical Ranking of the Three Classes of Scrap.

4.4.2 Industrial Recycling

As shown by loops (2), (3), (2') and (3') in Figure 28, new and old scrap may be recycled by primary producers or secondary producers depending mainly on the material. For example, secondary aluminium, except for some of the highest quality scrap, is recovered by separate secondary smelters. Wastepaper ("paperstock") is not usually reprocessed by primary paper mills, which, however, recycle their own "broke". Copper scrap is processed by primary or secondary producers depending on the basis of impurity content and location. Ferrous scrap is consumed by steel mills and foundries, which in this connection may be considered primary producers. The small but growing amounts of glass recycled are usually processed by plants which primarily use virgin raw materials.

Some manufacturing industries absorb new (processing) scrap and old scrap as shown by Loops (2a) and (3a) in Figure 28. This is particularly true of brass and aluminium. The technical and economic advantages can readily be seen, but the quantitative aspects are not well known because of the lack of pertinent statistics.

The secondary materials industries handle a large fraction of all recycled material originating in the manufacturing and post-user stages, especially metal scrap and paperstock.

Much secondary material is recycled by general firms, but some types are handled by specialists. Examples of the latter are automobile shredders and shipbreakers. In the United States an interesting development is the closed-loop recycling of aluminium cans, which yields new can stock from discarded cans. Similarly, old newspapers may be recycled into newsprint.

4.4.3 The Contributions of Industrial Recycling to Supplies

The main significance of established industrial recycling derives from its contributions to the supply of materials. These contributions can be analyzed quantitatively, but several conceptual points must be clarified first.

The ratio of actually recycled to potentially available or "recyclable" material (the "recycling intensity") is an important measure from several viewpoints – including those of conservationists and environmentalists. However, the amount of material available in any time period can only be estimated and for this reason the ratio has limited usefulness in spite of its appeal as an index of performance. A more useful measure, which will be considered below, is the ratio of recycled to total material consumed.

The distinction between the three classes of recycled material should be taken into account in a statistical analysis. Home scrap is a special feature of primary production that is of engineering rather than economic interest and is of little concern to outsiders. For most materials, information on the volume of home scrap produced and consumed is not readily available. Steel is an exception – published statistics show that in recent years 50 percent of total steel production in the UK and the US was derived from scrap; in the US at least half of this was home scrap, less than one-quarter new and somewhat over one-quarter old scrap.

New scrap poses an interesting question of interpretation: Does it constitute a true contribution to supplies– From a physical viewpoint new scrap increases the total flow of scrap through primary and/or secondary production facilities. In economic terms new scrap increases the volume of scrap entering the market. Also, if new scrap were not recycled, the demand for material from other sources would increase. From a fundamental resource standpoint, however, new scrap is a stock item which, except for processing losses, circulates in the combined production and manufacturing systems.

New scrap has another interesting characteristic. Since it depends on the volume of production of manufactured goods, it maintains a steady level under stable business conditions. However, when business conditions change, the volume of new scrap lags behind: it fails to keep up with a business expansion and thus does not immediately alleviate a possible scrap shortage and during a contraction it adds to the amount of overhanging supply.

The availability of old scrap depends on the life cycles of the goods from which a material is recovered, the volume of these goods produced at the beginning of their life cycles and the ratio of actually recycled to potentially recyclable material. This ratio, i.e., the "recycling intensity" was already mentioned earlier. The corresponding relation is shown schematically in Figure 30.

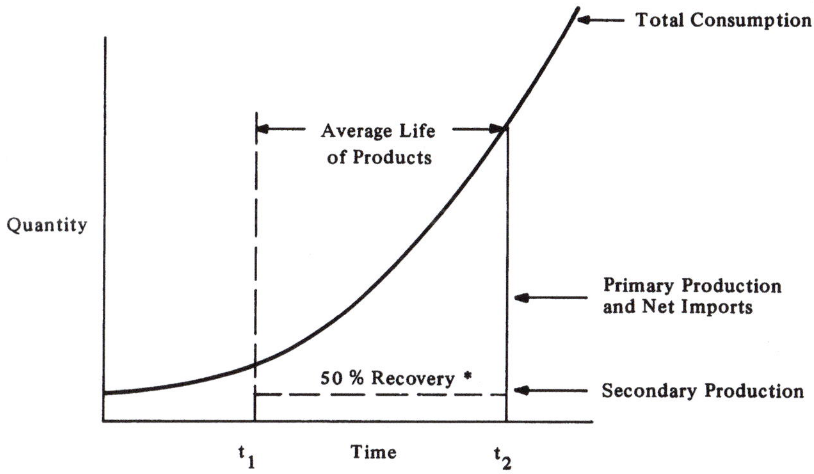

*Note: 50% recovery refers to metal produced at t_1 which results in only about a 15% contribution to consumption at t_2.

Figure 30: Relation between the Average Life of Products, Secondary Production and Total Consumption of Materials.

The diagram explains that even with high recycling intensities the contribution of old scrap may be only a small fraction of current consumption. Moreover, because of an increase in consumption the fraction may actually decrease; in such a case the decrease does not necessarily indicate a deterioration of the recycling performance.

Figure 30 indicates the basic pattern of exponential growth of material consumption, prevailing in many highly developed countries until the mid-70's. Let's take the example of aluminium. Its total consumption (primary and secondary) was doubling every decade since the end of World War II and primary production was growing faster (9% p.a.) than secondary (5–6% p.a.). The average lifetime of a product made from aluminium was 10–12 years. But now this is changing, because primary production may grow only 4 or 5% p.a. and the average lifetime may go down as recycled cans become a major item (today, about 20% of total aluminium consumption goes into cans and canlids in the USA.)

4.4.4 Resource Recovery from Municipal Solid Waste

The current development of "resource recovery" from municipal solid waste (MSW) is introducing a new factor into the materials production-consumption system. MSW is a more diverse mixture than most other waste materials. Moreover, its components have low value. On the other hand, the large volume of MSW accruing in densely populated areas may yield economies of scale in the separation process. Voluntary or mandatory separation at the household level ("source separation") under favorable circumstances also holds some promise.

A major task facing resource recovery from MSW is the development of steady markets. Under current conditions, the recovery of energy from the combustible fraction of MSW is the chief potential source of revenue, followed by ferrous and possibly aluminium scrap.

Resource recovery from MSW illustrates the systems nature of recycling. The reduction in the volume of material requiring disposal is an important item in the economics of resource recovery.

4.5 Assessment of Recovery Technologies

For resource recovery from MSW, two main features should be realized: the separation process can either be carried out at the front-end or at the back-end of the recovery plant. The latter means that, after combustion, certain materials can be separated magnetically or by other methods from the slags. The really interesting potential and technological challenge lies in front-end separation where the shredded waste is separated by floatation, magnetic and other processes. This has given rise to the development of new separation techniques. Similar technologies are needed and partly developed to separate various materials from automobile shredders.

A difficulty that arises is what to do with the organic components of the waste stream. They can be either burned or used for composting or animal feedstock, or as a raw material for various chemical processes. Another possibility is downgrading for instance plastic wastes into asphalt and other organic materials of secondary quality requirements.

The final target is the maximum recovery of all values in the "urban ore".

However, there are two major problem areas. A main technological problem is a consistent product quality. The other problem is the geographical location of the waste products in relation to further use. It is a major problem to utilize large quantities of glass, metals, paper and plastic, at great distances from locations where they are normally needed as raw materials input. This in turn means that unconventional uses must be found for many of the recovered products, for instance glass that can not be shipped to a distant glass bottle plant. Paper must be used locally either as a fuel, for thermal insulating materials, or perhaps for conversion into oil substitutes, which are still developmental. This brings us to the areas for research and development.

Many basic systems for reclaiming materials from MSW have been conceived but the proposed technologies are very expensive. One example is heavy media separation (the use of liquids with a density above 1.5 up to 3) for the separation of mixed plastics, mixed metals and glass.

Among the research proposals dealing with organic wastes, pyrolisis has been a favorite. However, being like most other proposals, this is a question of economics. As long as primary materials and primary fuels are relatively cheap, the largest quantity by far of MSW will be handled either by landfill or via incineration plants. Up to now, for economic reasons, far more than 90% of all MSW in highly developed countries will not go through a front-end separation system, because it is still too expensive. On the other hand, the "urban ore" represents an enormous value, but , only if these dispersed materials can be recycled technologically and economically.

4.6 Materials Conservation and Economics

We will restrict ourselves to a few views about this subject and quote figures collected in 1974 by the US Bureau of Mines for the USA.*

In Figure 31, a flow chart of the total mineral supply and recycling system for the US economy is presented (1974). It includes energy materials which are statistically difficult to separate from non-energy materials, keeping in mind, that oil derivatives are partly used to make plastics and further, that the materials industry is a large consumer of energy.

It is interesting to see that of the 1400 billion dollar net input into the product consuming sector of the economy, only 79 billion dollars (less than 6%) was the input for raw materials, including energy materials. An estimate shows that the energy materials account for about half of the value, therefore, the total value of the non-energy raw materials input into the product consuming sector of the economy was only on the order of 3%. Further, it is impressive to see that of $ 79 billion total raw materials input, only $ 4 billion come from reclaimed materials. Keeping in mind that half of the raw materials input was energy materials, we conclude that the value of the reclaimed secondary non-fuel materials is on the order of 10% of the corresponding primary raw materials value ($ 4 billion).

* See Bureau of Mines Information Circular / 1976, IC 8711, "Recycling Trends in the United States: A Review"

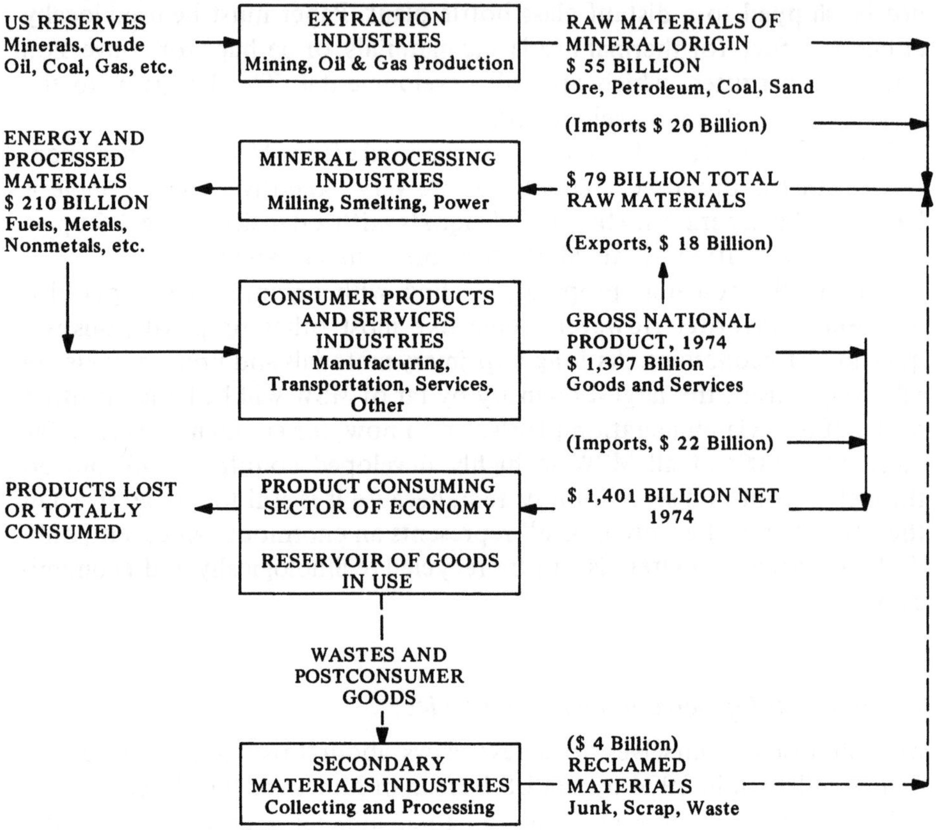

Figure 31: Mineral Supply and Recycling Systems in the US Economy, Estimated Values for 1974.

While over the last three decades there has been an enormous increase in US mineral demands, neither a resource constraint nor the cost of materials is a main driving force to increase resource recovery from MSW. The main reason so far is limited space for landfill disposal of urban waste and/or pollution of ground water by leaching of toxic impurities from waste deposits.

The total solid waste generated in the USA, in 1974, was estimated to be around 4.7 billion tons of which half was of agricultural origin. The mineral and metals industries waste was around 1.8 billion tons. Municipal, urban, industrial, solid waste accounted for only 300 million tons. This figure includes garbage, trash, demolition-debris and fly ash from power generating plants, but excludes sewage sludge. Of all 80 mineral industries, only 8 generated 80% of the total mineral wastes.

The copper industry generated the greatest amount of waste by its ore tailings, the iron and steel industry ranked second, coal and phosphate rock producers were the third and fourth, respectively. Most of these mineral wastes generated by the mining and milling industry are very large in quantity and very low in value. Logically, the only incentive to deal with these low value wastes is protection of the environment and especially water.

In brief contrast, it can be stated that at the manufacturing and consumer end, recycling has great economic importance because the US recycling industries now supply more than 25% of all aluminium and zinc consumed, 40% of the copper, 45% of the iron and 50% or more of the lead. Very little of it comes from MSW refuse.

4.7 Longer Lifetime of Consumer Goods

The drive for products with a longer useful life has been described often in mass media but may have its practical limits.

In West Germany, for instance, a car manufacturer has developed a long-life automobile with an expected or even guaranteed lifetime of 25 years where certain parts of the car have to be exchanged after 12 or 15 years. If the reader would imagine that he would use a car today which is more than 20 years old, it is easy to see, that this car could be outdated by insufficient safety devices and other features which the consumer expects in his car. Further, the wide-spread use of outdated technology would slow down technological development in general.

Therefore, the practicality of the idea is questionable from different angles. This is why the postulate for longer lifetime of consumer goods is a two-sided coin. Certainly, a gradual and steady change in this direction is desirable. By no means is the return to planned obsolescence feasible.

Therefore, a certain happy medium regarding lifetime of products and definitively less and less materials use per unit of output for many consumer products will result.

4.7.1 Redesigning for Recyclability

Redesigning, in the future, may be an important development for materials intensive consumer goods like refrigerators and other household machines and mainly automobiles. The goal obviously will be to have a design, where the main components could easily be dismantled for separate reuse or remelting.

Many details remain unresolved. For instance: if the wiring harness of cars could be changed from copper to aluminium, the steel scrap would be easier to recycle and higher in value, because copper together with other impurities are detrimental to most steels. If all aluminium sheet and extrusions on a vehicle would be of an "uni-alloy" and another one for all casting, this would enormously facilitate recycling aluminium from old automobiles. Metals from shredding operations are rather ill-defined in composition and present problems for further use. Dismantling a car is labor intensive, therefore what may be needed is the development of a robot to do this job.

Similar problems exist with other products, eligible for more recycling. For instance aluminium can bodies and lids are of two different alloys – again an uni-alloy could be used and simplify the recycling problem.

4.7.2 Remanufacturing and Reuse

Already today certain metallic components are reused on a large scale. This is true for locomotives, agricultural machines like tractors, automotive engines and small or medium electric generators or motors.

The incentive was clearly economic. Studies of the OTA and other institutions indicate however, that a wide field for remanufacturing is still untapped up today. Certainly, reuse is much more economic than remelting, especially when energy accounting enters the considerations. A first necessary step is redesign of many products, which are today discarded after 5 or 15 years of use (like household appliances) to make reuse of components easier and economical.

5 Materials and the Automobile Industry

The automobile industry is a large user of materials. Their pattern of materials use will change in the future, particularly in the USA. A primary reason is US government mandated fuel economy requirements. In simple text, this means a reduction in the weight of cars produced in the USA. The automobile industry in other countries is not pressed so hard in this respect because the cars are already lighter and smaller than their USA counter parts. Therefore, in this section, the discussion will center on the US automobile industry.

In the United States, the redesign for lighter weight is taking two major forms – materials substitution and downsizing. In materials substitution, the maximum weight savings will depend upon the particular application and the critical physical property required of the part. Strength, durability and quality cannot be sacrificed when a substitution is made. In downsizing, the industry is concerned about providing an acceptable product for the consumer. For the North American market, downsizing alone will not achieve the weight reduction required since driving habits and needs are different than elsewhere. Therefore, in order to meet regulations certain adjustments will have to be made by both the industry and the consumer.

This means the consumer will have to adjust to a somewhat smaller car and the car industry will have to use materials in new ways. But, using more or less of a material is not as simple as it may sound.

The supply issue is of great importance. Availability is a genuine concern since the automobile industry currently consumes 68% of all the lead used in the USA, 33% of all the zinc, 20% of the steel, and 9% of the aluminium.

Let us now take a look at the 1977 material mix in an average US car and see what changes are projected for 1985 (Table XXXVIII).

Table XXXVIII: Material Mix of Average Car

	1977		1985	
Material	%	kg	%	kg
Cast Iron	17	290	10	118
Steel: high strength	3	50	18	210
non-high strength	59	1002	41	484
Aluminium	3	50	8	95
Plastics	5	86	10	118
Other	13	222	13	154
Total Vehicle Dry Weight		1700		1179
Total Vehicle Inertia Weight*		1905		1361

* Includes fluids, weights and passengers.

Source: Ford Motor Company, December, 1978.

From Table XXXIII it can be seen that the 1985 automobile will not use a lot of exotic materials. There will be an increase in the amount of aluminium, high strength steel (HSLA) and plastics.

5.1 Materials Substitution Possibilities

Aluminium: Aluminium is one of the key materials in the substitution program. Because it can be used in place of cast iron, it offers the possibility of direct substitution for cast iron engine components. When the principal design criterion is strength and not modulus, one kilogram of aluminium replaces roughly two kilograms of steel or cast iron.

Die-cast parts will probably account for the lion's share of the usage. When the weight savings per component is considered, it is obvious why increasing amounts of aluminium will be used. For example, an aluminium intake manifold can save 11 to 14 kg per car. An aluminium cylinder head can take 14 kilograms off the total vehicle weight. Brake drums and axle housings are other candidates for aluminium.

Stamped sheet aluminium offers a significant weight-saving potential, but there are cost penalties in the most applications – with the penalty being as much as 80% more than conventional steel in some cases.

The very fact that new material applications are considered even at a cost penalty is typical of what is happening in the automobile industry. Traditionally cost has been the key factor in a materials substitution decision. Today, the guideline is cost per kilogram of weight saved. If a proposed substitution looks good from that viewpoint, it will be considered for production. Aluminium hoods, deck lids, doors and fenders are sheet aluminium parts that are typical examples of wrought aluminium use.

High Strength Steel: One of the most attractive alternatives is high strength steel, although there are some technical issues involved in the substitution of high strength steel for conventional steel. Formability decreases with increasing strength. Weldability is affected by the carbon and alloy content. Corrosion protection requires special consideration. Also, cost penalties could be involved in some cases. However major obstacles are not foreseen, so numerous applications can be anticipated.

Plastics: Compared to high strength steel, the next highest growth of all automotive materials will probably be in plastics.

A number of new applications already have been suggested. One is seats, including the springs and frames – totally plastic seats. Plastic fuel tanks appear to have potential and plastic hoods, deck lids and doors are the subjects of exhaustive industry development programs in the United States and other countries. Such components could result in a 50% weight saving. Hybrid panels, consisting of an aluminium or steel skin and a plastic core, are also being considered.

These considerations are possible because of advancements that have been made in structural plastics technology. For example a thermoplastic stamping process now allows the stamping of plastic parts at rates exceeding those of any other known plastics process – up to 500 an hour for small parts and up to 250 per hour for larger more complex parts.

5.2 Materials Substitution Problems

Aluminium: Recycling of materials is an issue which must be dealt with as substantial materials substitution occurs. Ways must be found to efficiently recover these materials, because if they can not be recovered it could cause problems in the overall materials cycle. Some of the anticipated problems are described in the following paragraphs.

Technical problems in the recovery of aluminium alloys from future automobiles could arise from the diversity of compositions. The alloy developments of the aluminium companies, the preferences of automobile manufacturers, and the mechanical requirements of the applications of wrought alloys mean that a range of compositions will probably be encountered in the aluminium alloys recovered in shredders. Without effective technology for refining secondary aluminium (except for the removal of magnesium) difficulties in recycling aluminium alloys can be foreseen. There is a question whether it will be economical to remove large, wrought aluminium alloy components such as hoods and trunk lids in a special disassembly operation.

Another crucial question concerns the size of the automotive and other markets for aluminium alloy castings made from secondary aluminium. If these markets grow sufficiently large they will be able to absorb increased amounts of scrap of both casting and wrought alloys. However, if markets for secondary casting alloys fail to grow sufficiently, effective methods for producing wrought alloys from automotive scrap will be needed.

High Strength Steel: The expected large increase in the ratio of high strength steels to other ferrous metals in future automobiles will create a substantial technical problem because of the alloying elements present. Elements such as vanadium and titanium which are not recovered in the usual steelmaking processes will be lost and part of the chromium and manganese will also be lost. On the other hand, some high strength steels contain alloying elements such as molybdenum and nickel which are not eliminated in steelmaking; introducing them into steels, through scrap utilization, is undesirable because they affect the formability and heat treating behavior of steels in an unpredictable manner. No metallurgical solutions to these problems are known or likely to be found. The increase in the concentration of residual elements in steel in all probability, will be one of the most adverse effects caused by changes in automotive materials between now and 1985.

Plastics: The increased use of plastics in automobiles can cause major technical and environmental problems. Plastics are difficult to separate from other components and from one another.

Hybrid panels consisting of a steel or aluminium skin and a plastic core are being considered for automotive applications. For recycling these components, the metal would have to be separated from the plastic. Shredders could perhaps accomplish this separation but this would depend on the method of bonding the skin and the core together. And, there is no published indication that this has been recognized as a problem. In general, recycling of plastics from discarded automobiles is not on a scale adequate to assure resource conservation and waste disposal.

The projected changes in the materials mix of the 1985 automobile could have an effect on its overall recyclability as well as on the technical and organizational features of the recycling system. Thus, the automobile and associated industries are faced not only with an assessment of materials substitution and suitability but also an assessment of the recyclability of the materials involved.

6 Materials in Packaging

The purpose of packaging is to protect a product on its way from the producer to the user, so that it will not be damaged during transportation or storage.

The various end uses call for different types of packaging, covering a wide spectrum of materials and shapes.

To illustrate the size of the packaging business, let us have a closer look at the US packaging industry.

The total sales of the packaging industry in 1976 was nearly 33 billion dollars, which is about 3% of the value of all finished goods sold in the USA. It includes all types of rigid and semirigid containers, flexible packaging, caps, closures, etc.

The main end uses are for the food and beverage industry (51% of total sale of packaging), industrial products (24%), general products (13%) and consumer chemicals (12%).

The main materials used for packaging, in the USA, are given in Table XXXIX.

Table XXXIX: Main Packaging Materials in the USA

Material	% of Total Sales
Paperboard	33.5
Metals	27.5
Plastics	12.0
Paper	11.9
Glass	9.8

Many other auxiliary materials are used by the packaging industry for joining, fastening, glueing, as well as for surface protection, such as lacquers, paints and the like.

In the following pages, the most important materials used by the packaging industry are reviewed.

6.1 The Main Groups of Materials

Paper and paperboard: Paper packaging includes bags of all types used for grocery and merchandise, heavy duty bags, flexible packaging, paper and paper-foil wraps, lables, tags, tapes, etc.

Paperboard packaging includes corrugated containers, folding cartons, food containers, mil and beverage cartons, drums, and various rigid boxes.

Paper and paperboard are made from wood pulp and other cellulosic pulps.

Paperboard differs from paper, only by its greater thickness and weight. Most of it is made from corrugated sheets glued to one or more liners. It can be made exceptionally strong, and therefore paperboard finds wide use for light-weight, heavy-duty containers (folding cartons, food containers etc.).

Metals: The USA used in 1976 approximately 6.6 million tons of steel and nearly 1 million tons of aluminium for packaging. This represents approximately 7.5% of the total production of steel mill products, and 19% of total aluminium production.

Metal is used in cans, aluminium foil containers, aerosol cans, collapsible tubes, pails and drums, etc.

- Metal cans are used to preserve food for a long time. They have outstanding strength and durability, and resist heat, cold, moisture, etc. Tin plated steel cans dominated the market until the 1950's. Aluminium then made inroads into the can industry. In the USA, today, approximately 27% of all metal cans are made of aluminium. Recently, tin-free steel cans were introduced. Easy opening aluminium lids are widely used today, especially for beer and soft drink cans.
- Aerosol cans are pressurized containers, used for deodorants, hair spray, paints, sprays etc. Most of them are made from tinplate. However, in Europe, aluminium aerosol cans are widely used.
- Aluminium foil containers are semirigid products, used in baking, or for processing and preserving frozen food.
  They have excellent barrier properties and are very compatible with food since they are sterile, nontoxic, and odorless.
- Metal collapsible tubes are used mainly for packaging of semiliquid products, such as tooth paste, food, cosmetics, medicines, etc. Most of the collapsible tubes are made of aluminium.
- Metal pails and drums are heavy-duty containers, used to transport liquids, chemicals, petroleum products and the like. They are made of steel, or tinplate, and characterized by their high strength and durability.
- Beer barrels are made from aluminium or stainless steel.

Steel is the dominant metal used in the production of metallic containers. Steel sheet for food cans is used mostly in gages between 0.15–0.3 mm. The tinplating process is applied to protect the steel from corrosion. Originally, the steel sheet was passed through a bath of molten tin. More recently, electrolytic methods are preferred. The thickness of the tin coating is up to 0.002 mm.
Plain steel sheet or tinplated sheet may also be coated with organic coatings or enamel, to provide corrosion protection for the inside and outside of cans.

Aluminium is penetrating the packaging sector on an increased scale. It is used in can making, semirigid containers, and in flexible packages. Aluminium sheet approximately 0.25 mm thick is used for can making and the manufacture of beer and beverage cans. Unlike steel, aluminium for packaging purposes can be further rolled into thinstrip of 0.02 mm or into foil, down to thickness of 0.005 mm.

Aluminium foil is impermeable to light, gas, moisture, odor, and solvents. It has stiffness and dead-fold characteristics, not found in any other flexible packaging material. It is therefore used considerably for packaging a wide variety of processed food.

In addition aluminium is used in many applications as a laminate in combination with various plastic films.

Plastics: It is estimated that in 1976, approximately 2.7 million tons of plastic materials were used for packaging (which is one-fifth of the total plastic consumed in the USA). Of this, one-third was used as rigid and semi-rigid containers and over one-third as film.

Plastic is a most suitable material. It is easily formed to shape, or used for wrapping food products as a heat shrinking or stretching film.

The most widely used plastic materials are polyethylene, polypropylene polyvinylchloride and polystyrol.

Besides the use of a single protective plastic layer, composite (laminated materials), were introduced, such as paper or aluminium foil laminated to a plastic sheet. Composite materials are widely used today for packaging by the food industry.

Plastic bottles and plastic squeeze tubes are the most common packaging products.

- Plastic bottles and plastic squeeze tubes are the most common packaging are generally resistant to water, oil, chemical and gases, but show permeability to oxygen and vapor.
  They are used for packaging household chemicals, cosmetic and toiletries, food and beverages, and medical and health products.

– Plastic squeeze tubes compete with metallic collapsible tubes. Their advantage is that they retain their shape and thus keep their esthetic appeal. They are mostly used for packaging cosmetics and toiletry products.

Various other packaging items are found in today's markets, such as closures, flexible packaging, cushioning materials, trays, blister packs. A relatively new packaging product is the plastic soft drink can which has been recently introduced into the market.

Glass: Glass is one of the oldest known materials, produced already around 7000 B.C. In Egypt, in 1550 B.C., an important bottle making industry developed.

It is an excellent packaging material, mainly due to its transparency and chemical inertness to the packed product. Its main disadvantage is its tendency to break and shatter.

The raw materials for glass manufacture are sand (75%), soda (15%), and limestone or chalk (10%). The components are heated in a gas or oil fired furnace to a melting temperature of 1400 to 1500° C.

The molten glass is formed to the required shape by various operations, i.e., casting, drawing, molding, blowing etc. Equipment of medium or high automation is used.

After shaping it into products, such as glass bottles, a controlled slow cooling is carried out to keep the glass as free as possible from stresses.

Different glasses have different mechanical, optical, thermal and chemical properties. Glass has relatively good compressive strength, but low impact resistance, which is its major drawback.

Transparency, is one of the great assets of glass, and an important marketing feature. On the other hand, passing ultraviolet light might cause food deterioration. Colored glass is therefore used, as in the case of beer bottles, to reduce such effects.

6.2 Assessment of Trends in Beverage Cans and Sterilizable Food Containers

To illustrate the changes in technology and materials substitution in the packaging field we have chosen two typical cases:

– beer and beverage cans;
– sterilizable semirigid packaging.

In both cases, the packaging industry is full of competition, by substitution of materials and processes. Recently, energy aspects and public pressure to reduce litter and waste of materials, have been main forces for continuous change.

Beer and beverage cans: The typical sizes range between 0.2 and 1 liter. In the USA, Japan and several other countries, the 12 oz metal cans have a lion's share of the beer and beverage market. Various materials are in use: glass, plastics, laminated cardboard, tin plate, coated black plate, and aluminium. These containers are designed for either one-way-use (throw-away type) or for multiple use.

Only a decade ago, tin cans or more correctly, tin coated steel cans, dominated the beer and beverage market in North America. The first aluminium cans were made from slugs by impact extrusion, but this process did not yield a price competitive product. The aluminium industry then developed a tailor-made can sheet, which is ideal to be formed into a seamless can in two steps, by using the draw and iron (D & I) process: first a cup is deep drawn and then the wall of the cup is ironed to the final dimension. Today, these drawn and ironed cans are produced in highly automated lines at a speed of 400 to 600 cans per minute.

A typical can plant produces 400 million cans a year. There is also great effort ongoing by the steel industry to utilize black plate with special coatings for the D & I process. So far, the final solution had not been found.

In comparison to steel cans, aluminium beer and beverage cans have certain advantages: less weight, better corrosion resistance, and longer shelf life.

One feature which must be briefly mentioned is the easy-opening-lid which appeared about a decade ago and has been modified since then to arrive at a safe and easy opening of the can. Here, aluminium had a clear advantage over steel, but again, the steel industry claims that it can produce today an all-steel can, including an easy-opening-lid. But so far, no visible market penetration of a new, all-steel can with an easy-opening-lid has been seen.

It will be interesting to watch the "race" between different materials for the can market. Today, it seems that the great efforts of the aluminium industry to achieve a large recycling rate, came just in time, to keep aluminium in this market (further details see p. 182).

Sterilizable semirigid packaging: In the past, sterilized food was packed mostly in steel cans. Aluminium has been more and more competitive in this market over the past ten years. First, shallow aluminium cans appeared on the market for packaging of fish, soups and other food stuffs, mostly in small size cans. The protection of the aluminium against the food stuff was achieved either by anodizing and/or lacquering.

For some years now, two new types of semirigid sanitary packs have been used. They are sterilizable containers and heat sealable sterilizable pouches.

1) Sterilizable containers, e.g., Steralcon® containers, are made from a two-layer laminated material where at the inside, mostly polyethylene or polypropylene is laminated to aluminium of approximately 0.1 mm thickness. After filling, the containers are closed by heat-sealing the lid. Then, they are submitted to a short sterilization time which may last approximately 15 minutes yielding a product of excellent quality.

These Steralcon packs actually compete against deep-frozen food stuffs and food packed in steel cans. Their advantage is often in institutional feeding (canteens etc.), as they are easy to serve.

2) Heat sealable sterilizable pouches are a relatively new development. Their laminated structure is similar to Steralcon packs, except for the thickness of aluminium, which is only around 0.02 to 0.03 mm. Strong plastic films are often laminated on both sides of the aluminium foil. The heat sealable pouches have a large market in Japan, and more recently in the USA and Europe. Originally they were used for soups, but they are suitable for many other food products.

6.3 Recycling of Packaging Materials

The huge annual consumption of materials and energy used by the packaging industry explains the enormous efforts to increase recycling or reuse of packaging materials.

There are, however, certain limitations to reach almost total recycling. The most important ones are the following:

- dispersity of the used packs in household, recreation areas etc.;
- economic and technological constraints for the recycling of different types of glass not compatible to be jointly used in the glass making plant (for instance due to different composition of white, green and brown glass);
- steel cans have a low scrap value and therefore make recycling economically unattractive.

In the future, containers and wraps made from combustible materials, like plastic, cardboard or paper, would be probably burned to generate heat, instead of being recycled. In the case of plastics, this approach is called "precycling".

First, plastic materials are made from gas, oil derivates or biomass. Second, plastic containers are made and used . Third, the used containers are burned (for instance in city garbage) for the generation of steam, heat, or electricity.

* ® = Registered Trade Mark

The Specific Case of Beer and Beverage Cans: The case of one-way containers has reached enormous importance because of the pressure to avoid litter on one hand, and the interest of the materials industry to recover used containers, on the other.

The aluminium industry in North America and Japan has made a great step forward by setting up the whole system for recycling of used cans.

In North America, more than 1 million mt of aluminium per year are used for making cans and easy-opening-lids. Recycling reached huge dimensions since the mid-70's. It is estimated that in 1978, almost 50% of all aluminium cans, were recycled. It is expected that up to 80% of the aluminium cans will be recycled within a decade.

The packaging industry is very sensitive to increases in material costs. When recycled aluminium will account for 60–70% of the total metal required, and primary only 30–40%, an increase in primary aluminium price will lose its impact on total material cost.

In the USA, now there is pending legislation to have a rather high enforced deposit on beer and beverage containers (e.g., 5 cents per 12 oz can), so that most of the empties would be returned by the consumer. This clearly favors the use of aluminium cans, because aluminium scrap has a rather high value compared to steel scrap or used glass.

Because of the Deleany act, in the USA, all plastics, which dissolve minute traces of chemical compounds into the foodstuff or beverage, are banned. Therefore, PVC, Acryl and Nytril plastics have come under severe restrictions.

In the long run, the plastic bottle could have a considerable impact on packaging of liquids. There are now improved plastic bottles in the market, which keep the diffusion of carbon dioxide or oxygen through the wall of the plastic container at an extremely low level, for many months.

6.4 Summary and Outlook

In the field of the beer and beverage can market, and of sterilizable containers we can see a typical example, where market pull and environmental and energy consideration result in material substitution, and technological development of processes. Within one decade, all these forces have created drastic changes in these kind of containers. It is quite difficult to forecast what is going to happen in the coming years. New materials and technologies might already be in the "innovation pipeline". However, their acceptability, will be judged by criteria of low energy, low materials wastage, plus extreme standards for health protection.

Plastics, under consideration for the "precycling" concept, will be quite competitive against glass and probably even against metal containers.

Therefore, there is a permanent, ongoing substitution process. Recyclability, or the possibility of "precycling" might become a dominating factor, affecting the acceptance of a certain "pack" by the public and legislators.

Bibliography

J.G. *Albert,* H. Alter and J.F. Bernheisel, "The Economics of Resource Recovery from Municipal Solid Waste", Science, Vol. 183, 1974

The *American* Society of Mechanical Engineers, "Measurement of the Impacts of Materials Substitution – A Case Study in the Automobile Industry", December 1978

R.U. *Ayres* and S. Noble, "Materials Scarcity and Substitution", IRT – 302 – R. Draft, Oct. 1972

Battelle Memorial Institute, "A Study to Identify Opportunities for Increased Solid Waste Utilization", Columbus, Ohio, 1972

P.E. *Becker* and H.J. Pick, "Resource Implications of Material Waste in Engineering Manufacture", Resources Policy 1, 1975

Roger *Betts,* "Recycling Resources", Omega, The International Journal of Management Science, Vol. 1, 1973

S.L. *Blum,* "Tapping Resources in Municipal Solid Waste", Science, Vol. 191, 1976

Barry *Commoner,* "The Closing Circle", Jonathan Cape Ltd., London, 1972

W.D. *Compton,* "Materials Substitution in the Automotive Industry", American Society of Mechanical Engineers, Dec. 1978

Dr. A. *Decker,* "Energy Accounting of Steel", 9th International TNO Conference, Rotterdam, February 1976

Paul *Ehrlich* and John Holdren, "Environment, Vol. 14, No. 3, 1972

"The *Energy* Accounting of Materials, Products, Processes and Services", Proceedings, 9th International TNO Conference, Rotterdam, February 1976

R.G. *Griffin* and S. Sacharow, "Principles of Packaging Development", AVI Publishing Co., Inc., 1972

H.E. *Goeller* and A.M. Weinberg, "The Age of Substitutability", Science, Vol. 191, 1976

J.J. *Harwood,* "Recycling the Junk Car – A Case Study of the Automobile as a Renewable Resource", Materials and Society, Vol. 1, 1977

Library of Congress, Science Policy Research Division, Congressional Research Service, "Materials Policy Handbook: An Outline of Legislative Issues of Materials Research and Technological Application", Congress Serial C., US Government Printing Office, Washington, D.C., 1977

"*Mining* and Minerals Policy", 1977, Annual Report of the Secretary of the Interior, Washington, D.C., 1977

Dr. Paul H. *Müller,* "Aluminium and Energy", Shareholders Meeting, Swiss Aluminium Ltd, Zurich, April 1978

National Academy of Sciences, Committee on the Survey of Materials Science and Engineering (COSMAT), "Materials and Man's Needs: Materials Science and Engineering", National Academy of Science, Washington, D.C., 1977

National Academy of Sciences, Committee on Mineral Resources and the Environment (COMRATE), "Mineral Resources and the Environment", Washington, D.C., National Academy of Sciences, 1975

National Commission on Materials Policy, "Materials Needs and the Environment Today and Tomorrow", US Government Printing Office, Washington, D.C., 1973

Nato Science Committee Study Group, "Rational Use of Potentially Scarce Metals", NATO Scienific Affairs Division, Brussels, 1976

Petrochemical Handbook, "Hydrocarbon Processing", November 1973

David *Pimental* et al.: "Food Production and the Energy Crisis", Science 182, 1973

J.H. *Price*, "Dynamic Energy Analysis and Nuclear Power", in A.B. Lovins and J.H. Price "Non-Nuclear Futures", New York, Friends of the Earth / Ballantine 1976

James A. *Rauch*, "The Kline Guide to the Packaging Industry", Chas. H. Kline & Co., Inc., Fairfield, N.J. 1977

J.W. *Sawyer*, "Automobile Scrap Recycling: Processes, Prices and Prospects", John Hopkins Press Baltimore, 1974

W. *Seifritz*, Schweiz. Technische Zeitschrift Nr. 7/7, February 1977

Dr. W. *Sies*, "Probleme der Weltrohstoffwirtschaft am Beispiel der NE-Metalle", May 1978

John S. *Steinhart* and Carol E. Steinhart: "Energy Use in the US Food System", Science, 184, 1974

Alvin *Toffler*, "Future Shock", The Bodley Head, London, 1972

Ullmanns Encyklopädie der technischen Chemie, 4. Auflage, Band 7, 1973

US Bureau of Mines, "Energy Use Patterns for Metal Recycling", Information Circular 8781, US Government Printing Office, Washington D.C. 1978

US Bureau of Mines, "Recycling Trends in the United States: A Review", Information Circular 8711, US Government Printing Office, Washington D.C. 1976

US Department of Commerce, "Proceedings – Materials and National Issues Conference", Washington, D.C., November 18, 1977

V Research and Development Opportunities

By J.P. Clark* and F.R. Tuler**

A 1978 National Science Foundation (NSF) investigation found that no "key" scientific or technological breakthroughs have occurred recently or are imminent which would substantially ameliorate overall patterns of mineral usage or supply in the USA in the very short term. However, incremental R & D investments, if they are made soon and provide for continuity of effort, can produce supply benefits over the longer term.

Although the benefits of long-term basic research will not be available for at least ten years, continuity will supply a steady stream of results after that. R & D opportunities exist in each stage of the minerals/materials cycle.

1 Exploration and Discovery

In an effort to increase the available resource case, R & D can advance the understanding of mineral formation, the development of the geosciences and their applications to mineral exploration. Studies of the genesis, localization, and discovery of "blind ore bodies" that have no surface manifestations would also be appropriate. R & D is needed to foster new techniques for the discovery of ore bodes. Some examples of this are downhole sensing devices, and advances in instrumentation methods to learn more of the "signature" of complex natural material, as well as R & D concerning the sensing, information-processing and transmitting functions of earth orbiting satellites for exploration. New exploration techniques will help in the achievement of a geochemical survey of the earth's crust, providing a more comprehensive identification of materials in short supply. Contributions to exploration technology will come from basic investigations of "Pathfinder" and trace elements for identifying the location of ore boundaries and from basic research leading to improved knowledge of the physical, chemical, and mechanical properties of rocks and minerals. Further knowledge of ocean floor geology and of the technology for ex-

* Assistant Professor, Massachusetts Institute of Technology
** Visiting Scientist, Massachusetts Institute of Technology

ploring, evaluating, and recovering marine resources will also serve to augment mineral supplies.

2 Mining and Extraction

With the current emphasis on more glamorous high-technology research (i.e., solid state electronics, nuclear energy, space research), not enough R & D effort has gone into the preparative side of materials.

A general study of mining and mineral reclamation methods is needed. Specifically, R & D could be focused on the development of new equipment for mining, i.e., the use of plasma and rocket nozzle technologies for drilling hard iron-bearing taconites; techniques of *in situ* recovery and new reclamation technologies which would increase the efficiency of resource recovery and more cost effective methods for mining low-grade ores. The development of remote control technologies, and the development of "robotics" would increase mine production in hostile environments.

Advances in extraction technology are needed that are more efficient, less costly, consume less energy, and emit less pollutants for extracting basic materials from progressively lower grade ores. The development of new technologies is necessary to improve the physical efficiency of resource extraction. This requires research on electrostatic and electrophoretic methods of separation. Pertinent long-range research would study the simultaneous extraction of several materials from "ores" like granite.

Many opportunities for improvements in mining through advances from R & D lie in the area of mine design. More efficient extraction could be obtained through improved basic understanding of mechanical responses of rocks and minerals to temperature, pressure, and chemical environments.

A major problem in deep underground mines is control of mine environments. Increased attention must be given to more effective methods for cooling, ventilating, and dust control if increased productivity and workplace health and safety goals are to be achieved.

3 Mineral Processing

Research and development is needed to improve the efficiency of materials processes and to develop processes that allow us to economically recover resources which have heretofore been unused. Methods for processing lower grade ores and for increasing the percentage recovery from ores and scrap are needed.

The comminution of ores for the liberation of minerals is the most energy-consuming step in the production of many minerals. Research should be directed towards increasing the efficiency of energy utilization in comminution, towards finding ways of selectively liberating minerals and the development of more effective methods for fine grinding.

Increased mineral recovery requires the development of methods to separate and concentrate mineral values found in ultrafine size ranges. Other areas in mineral processing that require attention include the agglomeration of ores and of the chemical alteration of minerals before separation. Finally, the concept of by-product recovery should be stressed. Often by-products are not recovered, even though all of the mining and comminution costs have been incurred.

Research is needed to improve the technology for controlling emissions of associated elements in primary processing operations of most metals. Opportunities exist for increasing the efficiency of energy utilization in individual processing steps as well as in the total integrated system of related unit operations. Further, basic research in thermal chemistry and heat transfer is needed to improve calcining, roasting, and smelting operations.

4 Manufacturing Processing

Manufacturing processes can be the means of increasing material efficiency and lessening materials requirements by reducing the amount of starting material and the number of processing steps. This would not only decrease scrap and offal quantities but realize energy and money savings. A better understanding of the science involved in such heavy industrial processes as casting, rolling, forging, sintering, heat treating, etc. seems to offer opportunities for their ultimate improvement. Research in powder metallurgy forgings, various pressure forming methods, and laser technologies would be useful. Processing advances that lead directly from liquids and powders to the finished stages (by-passing the hot working and ingot stages) are needed.

5 Materials Design

New and improved materials are needed to permit engineering design that will reduce the amount of materials required to perform a given function. High performance materials such as superalloys, and some types of ceramics, plastics and glasses push the capabilities of the machinery which they comprise to high levels of achievement, while at the same time conserving materials. Composites are particularly attractive in achieving the dual goal of high performance and materials conservation.

Improved performance will be the result of modifying existing materials and developing new ones. In creating new materials it will be instrumental to look to those materials which will open up new resources (i.e., synthetic polymers) and to bring into further use materials which have so far evaded our efforts (i.e., titanium and the silicates).

Improved performance of materials will lead to better utilization. More must be known about fatigue, fracture, corrosion, flammability, thermal and photodegredation, creep, electromigration and electrochemical action and biological behavior.

A better understanding of material design will permit creative responses to design needs, generated out of an understanding of physical resources rather than empirical studies and design practices. This will lead to a better utilization of high performance materials. A research program to advance the fundamental understanding in materials should include: Studies of the relationship between microstructures and strength and ductility of materials; studies to bring to bear the techniques of applied mechanics and physical metallurgy on the behavior of important bulk materials; studies of the mechanisms of bonding between plastics, ceramics, and metals; and studies to predict fracture.

Conservation of materials can also be realized through improved performance and better utilization due to increased service life. To increase the service life of a material it is necessary first to determine the losses sustained from corrosion, fatigue, friction and wear, fracture, and high temperature service in different industries. Once these effects are known further research is necessary on ways to prevent them. Steps toward increased service life include work on corrosion in metals, and abatement of aging and deterioration in plastics. To understand and prevent corrosion, studies should be undertaken to determine and identify characteristics of new coatings suitable as protective media and to examine the possibilities of creating non-coating barriers between the material and the environment. Fatigue studies can contribute to lengthening the service life of materials by developing testing methods which can accurately predict long-term engineering service life based upon short duration loading properties, and by determining practical design methods for predicting fatigue life of completed structures and materials.

6 Production and Product Design

The integration of materials science and engineering with product design and manufacturing can lead to reductions in production costs and thus usage of materials.

A better understanding of the properties of materials and how to control them more efficiently will lead to better product design. Clarifying the functional requirements of a specific product will improve the design and realize savings in both costs and materials.

R & D is needed in the blending of material and product design. This blend can contribute to the conservation of materials by upgrading the strength and modulus of the material. This will allow for narrower safety margins and less material will be needed to realize a given function. Research is also necessary for modifying design theory and concepts. This could lead to the development of new ways to perform the same function, such as transparent fibers for transmitting telephone signals and new adhesives to replace rivets and nails.

Longer product and material lifetime may result from studies of life expectancies and the effects of corrosion and wear of the product. Studies should be undertaken to make products with a high resistance to corrosion and wear, easier to dismantle and separate for recovery and recycling. Systems design would also contribute to long life times and increased recycling of materials. Research to improve and develop non-destructive evaluation techniques, to further exploit predictive testing are other examples of studies in production and product design which could relieve problems of supplies and shortages.

Bibliography

Committee on Mineral Technology, Board on Mineral and Energy Resources Commission on Natural Resources, "Technological Innovation and Forces for Change in the Mineral Industry", National Academy of Sciences, Washington, D.C., 1978

"Federal Materials Research and Development: "Modernizing Institutions and Management", Report to Congress by the Comptroller General of the United States, December, 1975

D. *Gabor* et al, "Beyond the Age of Waste", Pergamon Press Ltd., Oxford, New York, 1978

Julius J. *Harwood,* Extracts from "Material Resources – R & D Response" in Requirements for Fulfilling a National Materials Policy, ed. by Franklin P. Huddle, Office of Technology Assessment, US Congress, 1974

Walter R. *Hibbard,* "Introduction to the Workshop on Substitution for the Symposium on Materials and the Development of Nations: The Role of Technology"

Dickinson *Hillman,* "R & D to Counter Materials Shortages Affecting Army Weapons Systems Development/Acquisition", in Proceedings of the Workshop on Government Policies and Programs Affecting Materials Availability, February, 1976

Robert A. *Huggins,* "Basic Research in Materials", in Science, Volume 191, No. 4228, 20 February, 1976

Priorities in Applied Research: An Initial Appraisal, National Academy of Engineering, Washington, D.C. 1970

S. Victor *Radcliffe,* "World Changes and Chances: Some New Perspectives of Materials", in Science, Volume 191, No. 4228, 20 February, 1976

VI Outlook

The era of *exponential growth* in materials consumption is drawing to a close. What will replace it? *Zero* or *no-growth* is not the answer. A policy of no-growth cannot work world-wide. There is a need to reshape attitudes about growth. Civilization is entering a transition period which will affect the materials industries. In this concluding chapter, we outline a few common problems facing the materials industries in the years ahead, and describe some key issues providing a birdseye view of the main trends which are visible today. Before doing this, we will briefly review the background for our analysis.

1 Resources for the Materials Industry

Nonrenewable resources, for the provision of energy and materials, were the backbone of industrial development until now.

No-one can deal with the issue of resources without defining his own position vis-a-vis "The Limits to Growth" study prepared for the Club of Rome, published in 1972, and subsequent studies with a similar theme. "The Limits to Growth" has been translated into 17 languages and about three million copies were sold within four years. The *limits issue* found its way into mass media all over the world. Although the slogan of a "finite world with finite resources" is well taken, some of the conclusions drawn are highly debatable and have given rise to plenty of misunderstanding. For example, that already in a few decades mankind would face a depletion of available resources. "The Limits to Growth" did not predict this. As a matter of fact, it very carefully said that this would be the result if existing trends were not changed. "The distinction between warning and prophecy was largely lost as, in the debate that followed, good theater drove out good analysis. The instinctive pessimists and the visceral optimists squared off. So much was obviously bad about both growth and no-growth that a credible diatribe against either could be derived from the same data base. The furor also brought out the analytical artillery of specialists who saw the concept of *limits* as helpful or hurtful to what they were advocating anyway. Environmentalists welcomed support for their warnings about resource depletion and environmental degradation. Development economists and business thinkers attacked the vulnerable methods of handling data in the *limits* analysis. Planners and leaders in developing nations were outraged by the notion that the world would run out of resources before they managed to get their share."*

The main misinterpretation was that the earth's crust does not contain sufficient non-fuel minerals and ores to sustain foreseeable needs in the next century. There are enough of such resources. However, this does not mean that materials can be wasted as they were in the past. Nor, does it mean that we should stop searching for new mineral sources. Each country should determine its own natural resources, and make use of them even if there are surpluses of specific resources elsewhere.

In the first part of this book, opposing views regarding limits-to-growth were pointed out. There, the difference between reserves and resources was explained and, that it is important to distinguish between resources, which have been identified, and reserves which can be economically and legally extracted at the time of intended use. It was also explained how reserves have grown over the years and how resources become reserves. A problem arises when the mass media confuses resources and reserves. Confusion also arises when the expected lifetime of either resources or reserves is calculated. In some publications a "static" lifetime is calculated, while others use a "dynamic" lifetime.

The static lifetime is a simple calculation. Known reserves are divided by the current rate of consumption which gives a number of years.

In simplified dynamic lifetime calculations, an estimated increase in consumption is taken into account. Obviously, such a calculation can drastically reduce the expected life of reserves, depending on the assumed increase in consumption. For example, if one were to assume that there would be sufficient ore from known deposits to last 10 000 years at today's consumption rates under a static calculation; applying a 5 per cent increase in consumption each year would create an exponential growth reducing the lifetime to 126 years with the dynamic calculation.

But, here a word of caution must be added. In assessing what the future holds, our past experience should be a guide. The so-called "dynamic" lifetime calculation does not consider that reserves, as well as scrap (which constitutes a resource) are also growing. Lifetimes for most materials have been increasing over the past 10 or 15 years because increased prices bring new reserves on stream, tend to curtail consumption and make recycling more attractive. In addition, the dynamic calculation does not take into consideration the reduction in use of basic materials per unit GNP in the main materials consuming countries and the ingenuity of man.

All of this has led to two major opposing schools of thought.

* H. Cleveland and T.W. Wilson, Jr., "Humangrowth", Aspen Institute for Humanistic Studies, 1978

Cornucopia versus Doomsday:* The polarization of experts into opposing groups creates an unfortunate situation. The public and even the materials industry does not really know what to believe. Two groups with extreme opposing statements are the cornucopians and the doomsday foretellers. The conflict between the two polarized groups of experts is characterized by thesis and antithesis. That is, either group can propose a thesis which can be countered by an antithesis from the other group.

Thesis and antithesis represent two extremes. The real situation is somewhere between the two, and therefore requires a synthesis. If one would like to develop a synthesis for a specific case, let us say iron ore supply to the West European steel industry, the synthesis will use more elements from cornucopia and perhaps none of the doomsday school at all, because iron ore is available in many different countries and regions. However, if one looks at the availability of manganese, cobalt and chromium for the steel industry in OECD countries, the synthesis may take some elements of the doomsday school and add substitution possibilities for these materials from the side of cornucopia.

The Paradigm: How can the coexistence of such opposing forecasts, like cornucopia versus doomsday, be explained? The answer is the "paradigm" of the expert or group of experts in question. A paradigm, as defined by Thomas Kuhn, is a set of broadly shared assumptions that governs scientific inquiry. According to Kuhn, a dominating paradigm of a certain period may be followed by a quite different one, because a watershed point is reached where the accumulated weight of discrepancies and anomalies that cannot be fitted into the old paradigm tips the balance, and scientists find it more profitable (in emotional as well as rational terms) to seek a new paradigm than to patch up the old.

Observation proves that, today, there are competing paradigms, of extreme polarization, simultaneously in existence. In forecasting, these paradigms are guided by a set of preconceived ideas. To quote W. J. Davis: "It is the paradigm that dictates which observation is important, which hypothesis merits attention, and which theory holds sway. Likewise, paradigms mold predictions about the future. In forecasting the future of industrialism, Bell operates by one paradigm, Heilbroner another. Bell explicitly assumes that energy and natural resources are effectively unlimited. Heilbroner assumes that energy and resources are limited and that democracy may be incapable of coping with the effects of their depletion."

Each significant paradigm obviously contains assumptions or even predictions regarding resource availability plus social-political developments.

* Cornucopia: The horn of plenty, represented as overflowing with goods, symbolizing prosperity and abundance. Experts close to cornucopia are H. Kahn and D. Bell. The doomsday school recruits its disciples from limits-to-growth groups.

A synthesis among leading paradigms should be provided by a well seasoned integrator, who is able to cope with technological inputs as well as those from the industry's total environment.

Reduced Demand for Materials: Let us pose once more the Club of Rome's original question: "Are there limits to the earth's supply of resources?". In analysing this question, one must take into consideration the slowdown in materials consumption growth rates in total.

Table XL shows that the weighted average growth rates of the last 25 years, world-wide, will be cut about in half for important primary metals during the last quarter of this century. The growth rate of plastics will be only about one-third of what it has been.

Table XL: World Demand for Metals: Weighted Average Annual Growth Rates (based on 5 year periods)

	1951/75 %	1976/2000 %
Crude Steel	5.4	2.5**
Aluminium	8.6	4.5**
Copper	4.4	2.8*
Zinc	4.6	2.9*
Plastics	13.0	4.5**

Sources:
* W. Malenbaum
** Various

Of course, there are great disparities in different world regions regarding both materials consumption per capita and growth rates. As stated earlier in the book, during economic development materials consumption first increases more rapidly than GNP. Perhaps, one could fear, that the growing population in the materials-hungry Third World could create a de facto demand which could not be met by the basic materials industry. But, as shown in Table XL, the total world de facto demand growth for key materials is now at rather modest rates. Reason: The industrialized world, until now the major consumer, is undergoing a slow-down in materials consumption, which compensates to a large extent for the higher growth rates in the developing countries. The main reasons for the reduced growth in materials consumption among industrialized nations are: pollution, energy, conservation, technological advances and societal changes.

Less dissipative use for instance of metals and more recycling will soon gather momentum. There has been a wide-spread misinterpretation of studies by the Club of Rome that dissipative use would waste a large part of common metals for ever. The real truth is in a way just the contrary. Through the activities of the basic materials industries as well as the consuming sector, the commonly used metals are being concentrated in the main consuming countries. For instance, consider lead, 100 years ago, lead was distributed in many ore bodies but now, it is available in the industrialized parts of the world for recycling and reuse in a concentrated form. The same can be said for other common metals like copper, iron and aluminium, where the non-dissipative use is now considerably increasing. In fact, "Spaceship Terra" doesn't lose any materials except very minute fractions which are sent away with satellites into space.

Renewable resources: Our remarks so far were related to key materials industries, using nonrenewable resources. However, we have to keep in mind that the volume (in cubic meters) of wood "harvested" per year is by far larger than the combined volume of all metals and plastics produced. Wood as an industrial material is gaining importance and more sophistication enters into the utilization of wood. Furthermore, wood successfully competes with plastics and metals in many everyday applications, for example, furniture and windows. Provision of wood is not energy intensive and with proper forest management ecologically sound. In the future, biomass (other fast growing plants besides wood) will have an increasing importance in providing energy through conversion into products such as "gasohol" in regions of the world with surplus arable land.

2 Materials and Ecology

Ecology and ecological constraints are terms very familiar to the materials industry. Today, ecology, environment and pollution are closely tied together and sometimes the words are used interchangeably. They are linked together with ethics, politics and regulations (environmental impact statements) and have altered industrial planning. Over the past ten years, these issues have become widely accepted as a major constituent of technology planning by the materials industry. This holds true for basic industries in both industrialized and developing countries.

A new problem which arises is potential over-regulation of industrial emissions that can lead to severe constraints resulting in an insufficient supply of materials. Some people propose *zero* pollution and still expect increased growth of basic material production. This is an impossible objective.

Even if man were not on earth there would be large quantities of what we call "pollution". Nature produces these large quantities. According to a recent paper by J. J. Mc Ketta, over 50 percent of the particulate matter (dirt) in the air comes from natural phenomena or processes. More than 60 percent of the sulfur dioxide in the air comes from nature. Over 75 percent of all the hydrocarbons in the air comes from nature. And, nature produces more than 93 percent of the carbon monoxide and over 98 percent of the oxides of nitrogen. Therefore, pollution problems must be attacked on a basis of what is known. Knowledge, not feelings can solve environmental problems.

The production and use of materials does create some "pollution". Efforts to control man made pollution are justifiable, but what are the costs? In the USA, it is estimated that each dollar the government spends on regulation requires private industry to spend 20 dollars. In many cases this means that money is spent to comply with a regulation, when the same money could possibly be spent to a greater advantage for the public in some other area.

This does not mean that we are advocating de-regulation or no controls. Mined-out areas should be reclaimed so that they can serve a useful purpose afterwards. Extraction of materials should not mean that harmful emissions are vented untreated into the atmosphere. Use of materials should not mean that old products, scrap, waste and litter are left lying around without attempts at recycling or reuse. In the past, materials industries often may not have paid enough attention to environmental issues. Nowadays, we have in many industrialized countries the opposite extreme: over-regulation. According to J. A. Overton Jr., President of the American Mining Congress, the total mining regulatory bill in the USA is $ 100 billion a year and rising. The US steel industry alone has to comply with 5600 different regulations.

In the end, of course, the public pays for compliance with over-regulation because the cost is passed on to them in the form of higher prices.

There must be a balance between regulations and industry. Trade-offs must be weighed. Technology is available to control pollution to reasonable levels. The materials industry, if not unduly hindered, will be able to perform its tasks because it has the:

- expertise and know-how;
- training to solve problems;
- flexibility to adapt to changing priorities.

The problem is to find a balance between industry and its most sensitive environment: the public and government. To de-emotionalize the dialogue, for each important case, a cost-benefit-risk analysis should be performed and made publicly known.

In the sector of mining and basic industries, the solutions for environmental issues are mostly "technological fixes", which are available at a price. In the sector of final products, or use of materials, there are societal changes. Consumers are becoming sensitized to avoid waste of materials and are actively engaged in recycling.

Among the societal changes, there are quite a few resulting in less materials input per unit of production or services. We cannot analyze all underlying causes for this trend. However, when looking into the future use of materials in highly developed countries, the important conclusion emerges that materials production and use, have a strong interrelationship with energy and/or information ("sophistication" or "software") applied to the use of the material. In fact, in numerous cases use of conventional materials, to achieve a given purpose, can be substituted by either energy or information.

3 Materials and Energy

The key issue for the materials producing and consuming industries, in the future, is the materials-energy relationship. Materials and energy are inextricably bound together. Industrial materials cannot be produced without energy, and energy cannot be produced and used without special materials. This interdependence has been increasing and holds true primarily for metals and common materials as well as for highly sophisticated materials.

Today, the energy costs for the production of primary metals range between 15% in the case of lead to 45% in the case of nickel (using lateritic ores). The total energy cost for primary aluminium is about in the middle of these two extremes. All these figures include previous refining steps in addition to smelting. If we would imagine that energy prices may double within a few years, these figures dramatically illustrate the impact of the rising energy cost component on the production of primary materials.

In this two-way relationship, between materials and energy, many important trade-offs exist. Typical examples are the use of leaner ores which in turn require more energy per unit of final material output. Another example is the "materials-energy issue" in design and use of automobiles.

An equally important aspect is the energy content of materials eligible for recycling. This can be shown in many different ways. In the precycling concept, paper, cardboard or plastic serve first, for instance as a packaging material and later on the energy content is used for generation of useful energy in solid waste incineration units.

Another aspect is the recycling of metals by remelting, as is now the case for all important common metals. Here, the initial energy required to extract the metal is to a large extent conserved through a small additional energy input.

In the provision of energy, materials play a very important role in construction of new energy production units as well as in the distribution and use of the energy. In steam power generation, a large variety of different materials are used for: boilers, steam pipes, turbines and condensers. The trend towards "supercritical" steam power generating plants with higher efficiency will put greater demands on materials.

Nuclear fission, in both thermal reactor systems and breeder reactors, subjects materials to other environments. Temperatures are high and radiation effects must be considered. Nuclear technology has introduced unique problems that arise primarily due to neutron bombardment of materials. These effects may cause shape changes in components and modify their mechanical properties. Radiation damage reduces the fracture toughness of steels and the degree of embrittlement is related to temperature, neutron dose and impurity content.

Magnetohydrodynamic energy conversion could enable coal and other fossil fuels to be converted into electricity without the need of rotating machinery. If the goal for an operating temperature of 2500°K can be met, efficiencies exceeding 50% have been predicted. So far, the system is materials limited and no materials are able to withstand the environment for more than a very short period.

Geothermal power causes materials problems mainly from the corrosion standpoint. Since each geothermal source has unique features, material selection is different for each installation. The main problems arise from the hot fluids and wear of drill rigs.

Energy storage systems are vital to planning an overall energy stragegy. Electrical energy storage is a prime concern. Storage batteries, mainly lead-acid cells, are used as a power source for some vehicles in commercial and industrial usage. Some power generating authorities use larger units as a means of storing energy for peak-load distribution. Many other battery systems are being investigated. These include: nickel-iron, nickel-hydrogen, nickel-zinc, nickel-cadmium, and sodium-sulphur. The design and manufacture of these cells is a very complex procedure in materials engineering.

Flywheels have long been recognized as a simple means of storing mechanical energy. They also offer the prospect of providing efficient storage of off-peak electrical energy to help electric utilities to meet peak load requirements. However, in the past, insufficient energy could be stored for a given flywheel weight to make the system cost effective. Now, with the development of fiber-composite materials, a flywheel energy storage system seems feasible. To take advantage of these fiber-composite materials, quite different concepts for design will be necessary. A flywheel that could store 10 to 20 MWh of electrical energy would be about 4 meters in diameter, with a rotational speed of 60 Hz and weigh between 100 and 200 tons.

Gas turbines have power-density and energy-density ratios that exceed those of all other energy converters. However, there is a continuing need for materials that will withstand even higher operating temperatures. A number of ceramics are among the candidate materials and include: silicon nitride, silicon carbide, and certain glass ceramics.

Solar power is widely regarded as the ultimate answer to the world's need for energy. Uncertainty and controversy arise over whether or not economic means of collecting and converting solar energy into useful forms can be developed.

This holds true for both the generation of heat and the direct production of electricity (photovoltaic conversion) on a large scale. In the case of solar heat, the operational time that is needed before the energy output of the system equals the energy required to produce the materials necessary for the solar collector system is particularly relevant. In many cases it will take years before the "harvested" net solar energy equals the energy used to make the solar equipment. For photovoltaic conversion of solar radiation into electricity using semiconductors, cost is the biggest factor. Ultra-pure, single crystals are very expensive. And, since a "solar farm" would have to be located in a hot arid region, utilization of the electrical energy would involve large storage and transmission networks. Here, enormous R & D efforts are underway to develop more cost-efficient materials systems.

Fusion power is being considered as a future energy source. A controlled fusion reactor's potential advantage over solar power is that it can be utilized any place on earth and is not climate dependent. The fusion reactor also has two major advantages over the fission reactor:

1) no radioactive wastes are produced from the fuel, and
2) the reactor is self-stopping in the event that an operational malfunction occurs.

However, major unresolved materials problems are an obstacle. These include:

- confining the hot plasma of several million degrees Kelvin, and
- radiation damage to structural materials and containment shields which will be greater than in fission reactors.

Candidate materials include stainless steel, nickel-based alloys and alloys of molybdenum, niobium, or vanadium – the last being advantageous because it does not form long-lived isotopes. The final choice will depend mainly on the operating temperature for the "blanket" surrounding the plasma vacuum chamber. Most likely, disposal problems will be quite difficult for the reactor structure subjected to the intense neutron flux as replacement becomes necessary.

There are more examples of the interplay between materials and energy production that could fill a book. Many challenges will confront scientists and engineers in providing synthetic fuels. Developing substitute fuels will involve huge, materials-intensive production units and almost every new or alternative primary energy source, or energy system will require improved or tailor-made materials.

In fact R & D expenditures today, with the objective to develop new materials or materials systems related to the provision, storage or use of new energy sources, are one of the best investments that can be made.

4 Sophisticated Use of Materials

Fundamental changes are taking place in materials usage in many industrialized countries. The trend is clear: less material is being used per unit of output or for a given product. Software and sophistication reduces the amount of materials required. This is explained schematically in Figure 32. In this figure, the amount of materials per unit of product is plotted against the amount of sophistication attached to the use of a given material.

The gravity center of several major materials producing industries has been indicated in this heuristic model together with a vector, shown as an arrow, to indicate what the trend in the next decade will be. Older industries, like steel and cement, have limited possibilities to move the gravity center of their product mix, within the next decade or so, considerably towards more sophisticated uses or drastically new technologies. For instance the basic steel industry will move only slowly out of its present position because the bulk of steel is used for rather simple products like railroad tracks or simple beams for construction. However, highly specialized steels like high strength, low alloy steel (HSLA), which are shown separately have a brighter future. The reader will remember that today HSLA is considered an important material for the automotive industry.

High temperature materials already have a lot of built-in sophistication as well as know-how in their final use. The opportunities for the high temperature resistant materials and younger and lightweight materials, like plastics, aluminium, titanium and magnesium, are more dynamic. It can be easily recognized that the automobile industry will move like most other

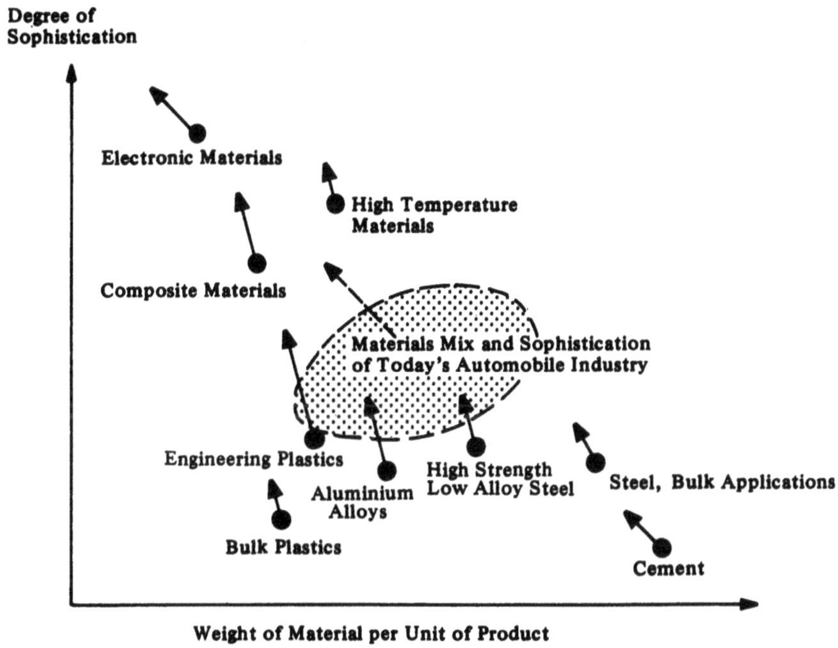

Note: Each point indicates schematically the gravity center of a given material (or group of materials) in the mid-seventies. The vectors show trends for the next decade or so. Degree of Sophistication is Know-how ("software") built into either the material and/or the final product.

Figure 32: Heuristic Model Providing Correlation Between the Quantity of Material in a Given Product and "Information" Attached to the Material and Its Use.

materials using industries, to the upper left of the heuristic model. For all vectors indicating innovative moves, R & D and new technologies will have a considerable impact within the next decade or so. An example would be the microprocessor industry, which would be placed around electronic materials in the upper left of the heuristic model.

In general, science and technology is the sector which creates new materials or composites as substitutes for previously used materials. Therefore, materials science and engineering will detect and develop new pathways into the future of materials using industries. It is true that science and technology are often blamed for the disadvantages of modern civilization like pollution, wasteful use of energy and the throw-away society. But it must be stressed that scientific accomplishment and resulting innovation are also very much needed to protect our society from undesirable developments which may even result in a decreased standard of living. R & D will be asked to provide new technological fixes to correct failures of the past.

Science, however, is dominated by specialists who have to be integrated into the total picture. Today, there are competitive claims on resource use. Complexity arises from priorities and trade-offs. No model can provide an optimum solution, when behavioral and social phenomena are included. The reason for this becomes obvious with a little thought because behavioral and social phenomena cannot be incorporated into a mathematical model. This being the case, how does one quantify human values?

The only possibility is through an integrated approach, which is an art, not a science.

5 An Integrated Approach

Integrating today's complex global problems is an evolving craft. It requires an integrator who is skillful in blending diverse disciplines, employing knowledge from a variety of sciences and economics, utilizing methods that are only partially formalized and teachable.

Integrators and an integrated approach are needed today because the materials industry cannot rely only on linear, incremental developments to cope with future challenges.

Figure 33 illustrates that in addition to the triad of raw materials, energy and ecology, the socio-political arena has strong interaction with the materials industry (and the triad as well)*. It is difficult for industrial management, specialists in industry, or academia to understand the socio-political pressures or opportunities. In fact, questions related to all 4 segments surrounding materials industry (shown in Figure 33) are much too important to be left to specialists with tunnel vision which inhibits broad, effective analysis and action.

Therefore, for a number of principal problems of the materials industry, specialists alone will be unable to provide the answers. In dealing with the problems illustrated in Figure 33, specialists normally arrive at suboptimal solutions, because they often work with a mono-causal concept. In searching for valid technology planning concepts for the materials industry, an overall picture is of utmost importance. In Harlan Cleveland's book, "The Future Executive", he quotes John Gardner:

"If we are to retain any command at all over our own future, the ablest people we have in every field must give thought to the largest problems of the nation . . . They don't have to be in government to do so. But they have to come out of the trenches of their own specialities and look at the whole battlefield."

* Although only 8 interactions are shown, it must be kept in mind that each sector interacts with all other sectors.

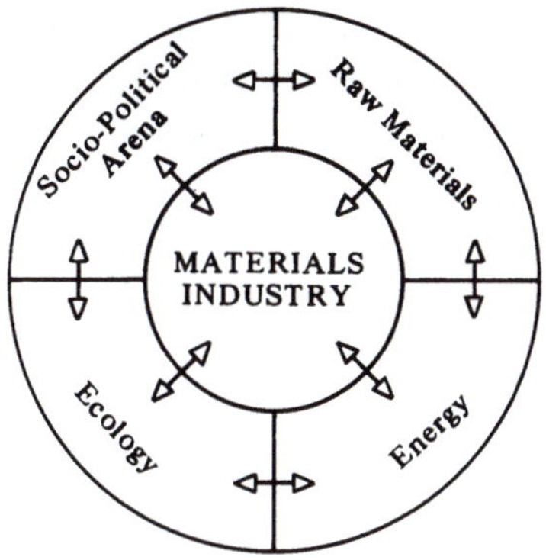

Figure 33: The Materials Industry and Its Interaction with Its Surroundings

This is particularly true for the materials industry today. It is essential to get an overall look at the entire "battlefield". Industry must know its position. At the beginning of this chapter (Table XL), it was shown that growth rates for the last quarter of this century will be greatly reduced. Yet, during the five year period 1970–74, world-wide investments were made for the production of energy and materials based on previous growth rates. This proves that drastic reductions in growth rates were not anticipated. How was it possible to make such a mistake? One of several possible answers is that too much reliance was placed on statistics and computer programs prepared by specialists for the decision-makers. General economic predictions are of little use today, although extrapolation of trends and previous growth was helpful in the past.

Now and in the foreseeable future careful analysis of the industry's total environment, as sketched in Figure 33, is a prerequisite for planning. Assessment of existing new technologies and recognition of their interaction with key issues, as outlined in Chapter IV, is a backbone for the planning process. In this book we have described, for selected materials industries, the criteria for technology planning and outlined opportunities for R & D. Proper integration of all these factors with the social-political arena is the only way in which solutions for the future will be found.

6 Trends and Issues

Today's events shape the future. We are living at a time when technological, economic, political, social and even religious developments play an increasingly important role for the energy and materials industries. These can have impacts on everyday life.

It is an obvious conclusion that the total materials cycle, from resources to the final product, is not only a key issue for modern civilization but will be subjected to many changes on a global scale over the next two decades.

For the future, a harmonious balance must be achieved between industry and society. Furthermore, the solutions to these problems have to be accomplished at a time when there is an uncertain energy future. To forecast what will happen in the automotive industry, in 1985, is not without guesswork. Lack of, or a high price for, liquid fuel could have an impact for a number of years.

Therefore, great difficulty arises in trying to project the future course of industries which deal with energy and materials. No-one can truly predict the future, but one can and should plan for the future, if trends emerge which have a high probability.

For the materials industry, there are numerous established trends and facts providing a sound basis for planning. A key trend for planning is "doing more with less".* Even today there are outstanding examples of doing more with less. A 200 ton jetliner can surpass the annual passenger-carrying capacity of the 85 000 ton Queen Elizabeth II. Similarly, a quarter-ton communications satellite can outperform 150 000 tons of transoceanic cables. Doing more with less primary material input will play a leading role in our future society. It will comprise measures and trends such as:

- recycling and reuse;
- conservation and substitution;
- miniaturization;
- using less materials-intensive systems.

The main driving forces for doing more with less are:

- basic materials will be more expensive;
- energy will be scarce and more costly;
- the developing world will need hardware;
- man's inborn desire to innovate and improve his skills.

As in the past, man's ingenuity will push the physical limits-to-growth back beyond the horizon again.

The Highly Developed Countries

A transition is taking place in the most highly developed countries. They are changing from industrial into postindustrial societies. A postindustrial society is organized around information, which is in contrast to an industrial society organized primarily around energy and the use of energy to

* A phrase coined by R. Buckminster Fuller already in 1961.

produce goods. This means that the emphasis is on services in postindustrial economies, rather than on material-intensive and energy-intensive production of goods.

Today, in the United States, more than half of the labor force is engaged in generating, processing, analysing, updating, communicating and distributing information. Information, which is derived from specific knowledge, becomes a basic resource in a postindustrial society. Treating information as a resource greatly alters thinking about the nature of economic growth. All other resources ultimately become dependent upon information and knowledge.

Changes for the materials industry are implicit in a postindustrial society. Innovation steps into the spotlight. More sophistication is attached to materials use and less primary material and energy is used per unit of output. However, during the greater part of the seventies, the basic materials industries in highly developed countries, especially the metallurgical industries, suffered from a pronounced recession which created a policy of retrenchment. This in turn limited the funds available for badly needed innovation.

Part of the problem stemmed from false judgement about future growth of these industries. But turning now to the future, it should be emphasized that solutions for the pending problems of the basic material industries can only succeed if the industries abandon defensive and adopt integrated, offensive forward strategies.

There are in fact quite a few opportunities for the materials industry. Findings in early 1979, by the US National Academy of Sciences, show that the rapid evolution of new materials that began largely during World War II is still under way. Advances should continue to be stimulated by the interaction of basic research and industrial technology.

Developments are expected during the next five years in eight, roughly grouped and partly overlapping classes of materials: metals and alloys, energy-related materials, information-related materials, polymers, ceramics and other inorganic materials, composite materials, renewable materials, and biomedical materials. These developments will illustrate the achievements and near-term potential of materials science and its application.

Conservation, recycling and substitution will assume greater importance. Recycling of materials, where it reduces costs, is expected to grow considerably in the next five years. Substitution of one material for another historically has been motivated by cost reduction, specific functional advantages, or supply considerations. The chief characteristic of a healthy climate for substitution is a stockpile of materials technology on which to draw.

The materials industry should enter a transition phase from a somewhat unbalanced into a more harmonious new system. It will include the coexistence of soft and hard technologies which will result in new cooperative relationships between industrialized and developing countries. New concepts will have to be adopted and it will be important to incorporate, into materials industry planning, the problems and opportunities of Third World countries.

The Developing Countries

Although there is no standard definition of a developing country, it is a country that falls below the standards by which "development" in some sense is measured. Obviously, therefore, where a country stands in development depends upon the scale that is used, and upon the ranking of that country on such a scale. Some countries may rank low on almost all credible scales. Others may actually be near developed status on some scales and behind on others.

Of particular interest to the materials industries are those countries not yet considered developed that have natural resources in the form of mineral deposits, untapped energy potential (such as hydropower) or energy resources such as fossil fuels. Others of obvious interest are those with large manpower bases which can provide the necessary transformation labor. Finally, these same large population countries are all potential market areas.

However, it must be kept in mind that industrialization requires a science and technology base. The main distinction between developed and developing countries is the ability to conduct and *use* research. Science creates technology which in turn creates new science which creates new technology and so on. This is a never ending spiral that does not exist in typical developing countries. Therefore, in the industrialization process, it is important not to cut the "umbilical cord" to the "science-technology spiral".

The rate at which development takes place in the developing world will depend very much on the success that is achieved in transferring relevant technology from the already developed countries, and on the success that is achieved in adapting this technology and providing "appropriate technology" uniquely relevant to developing countries.

Obviously many non-science and technology factors will greatly influence the speed and success of such adaptation and technology transfer. The nature of the development path that is chosen, politically and economically, will be important in terms of showing a credible and sustainable program for the future.

The degree of industrialization must be in line with available resources, existing social structures and the "ecological niche" of the specific country. Although most developing countries are at a very early growth stage, they can and should adapt to a growth pattern that is less demanding on nonrenewable resources, and take a lead in developing renewable resources. A well-publicized example is Brazil, using biomass not only for fuel but also as a feedstock for the chemical and plastics industry. Right at the beginning, the developing countries must emphasize a type of growth that is sustainable.

The soundness of the local political climate will be important to attract foreign investment. Investment provides income which can be channeled for development. One way of accomplishing this is through enlarged and improved cooperation with OECD countries. Another way is Technological Cooperation between Developing Countries (TCDC), where technology transfer is achieved through South-South cooperation. A third way to achieve progress is through projects in which several partners cooperate in a sensible way. Good examples are "tripartite consortia" where one partner provides the basic resource, another technology and/or marketing, and the third financing. Existing examples can be found in Africa, for instance, producing non-ferrous metals from the Red Sea and, bauxite-alumina from Guinean sources.

Such models of cooperation release new forces stimulating innovation which leads to economic and intellectual growth for all concerned. For materials industry development, in industrializing and industrialized countries as well, cooperation between nations will be one of the key issues in the coming transition.

In Conclusion

The coming transformation of the materials industries is a major element within a simultaneous conceptual revolution of the industrialized world that will have far-reaching consequences. Some knowledgeable observers have expressed concern over this change. Willis W. Harman, Associate Director, at the Stanford Research Institute, recently analyzed this subject and concludes:

"This overall transformation is proceeding with extreme rapidity, such that the most critical period will be passed through within a decade. Whether the social structure can withstand the strain is very much at issue, and this will greatly depend on how well we can understand the nature and necessity of the transformation while we are experiencing it."

However, undue pessimism is unwarranted. Despite the uncertainties ahead of us there are a number of clear-cut trends which can provide a basis for planning the direction of the materials industry. No matter how society changes in the future, the materials industry will always have an essential role.

At this point the reader may ask: "What then is the paradigm of the authors?" Although we are technologists, we realize that social, economic and technological change must be integrated. Our paradigm is that technology can provide the basic materials required for sustained growth and, further, can furnish any needed system for the use of materials. This includes materials conservation and improved ecological equilibrium. There are solutions to the problems that will confront us. We are convinced that man's ingenuity can solve these forthcoming problems through discovery, substitution, recycling and innovation. In this book we have tried to furnish some key information on these issues, on the foreseeable changes and, to arouse an awareness that the problems ahead of us must be tackled by systematic planning and assessment of impacts of alternative routes.

The dynamics of change will be dramatic for the materials industry. National governments and international bodies will play an increasingly important role in this change because issues are arising that can only be considered in a global context. Governmental policies of one nation on supply of energy or materials can have an impact on many other nations. Pollution from one country can produce "fall-out" in another country. Feasible solutions for these and many other issues will be quite different for the developing and industrialized countries in different parts of the world. There will be a growing *interdependence* among nations. The materials industry provides a most important link between industrialized and developing countries. Hopefully, both groups of countries will choose common ground on which to build new pathways into the future.

Bibliography

D.G. *Altenpohl*, "The Metal Industry in a Changing World", Materials and Society, Vol 3, No. 4, 1979

D. *Bell*, "The Coming of Post-Industrial Society", Basic Books, New York, 1973

H. *Brooks*, "Technology: Hope or Catastrophe?", Technology In Society, Vol 1, No. 1, 1979

H.B.G. *Casimir*, "Historical and Cultural Perspectives of Science and Technology in the Development Process". UNIDO ACAST Conference, Vienna, August 1979

H. *Cleveland*, "The Future Executive", Harper & Row, Publishers, New York, 1972

W.J. *Davis*, "The Seventh Year: Industrial Civilization in Transition", W.W. Norton & Company, New York, 1979

B. *Fritsch*, "Ein projektorientiertes, heuristisches Verfahren zur Modellierung von politisch, ökologisch relevanten globalen Zusammenhängen", Institut für Wirtschaftsforschung, Eidgenössische Technische Hochschule, Zurich, September 1976

B. *Fritsch* und G. Kirchgässner, "Zur Interdependenz von Wirtschaftswachstum, Energie- und Ressourcenverbrauch", Eidgenössische Technische Hochschule, Zurich, July 1976

D. *Gabor* et al, "Beyond the Age of Waste", Pergamon Press Ltd., Oxford, New York, 1978

W. W. *Harman*, "The Coming Transformation", The Futurist, February and April 1979

R. *Heilbroner*, "The Human Prospect", W.W. Norton, New York, 1974

Herman *Kahn* et al, "The Next 200 Years", William Morrow and Company Inc., New York, 1976

Herman *Kahn*, "World Economic Development 1979 and Beyond" Westview Press Inc., Boulder, CO, 1979

T.S. *Kuhn*, "The Structure of Scientific Revolutions", University of Chicago Press, Chicago, 1962

W. *Malenbaum*, "Anatomy of World Materials Demand: Its Role in Supply Availability", Mining Congress Journal, February 1979

D. H. *Meadows* et al, "The Limits to Growth", Universal Books, New York, 1972

Mining Symposium' 78 "Polities, Society, and the Minerals Industry" Materials and Society, Vol 3, No. 1, 1979

The *National* Academy of Sciences. "Science and Technology: A Five-Year Outlook", April 1979

National Science Foundation. "Preliminary Report on the Problem Analysis Phase for Task 8 of the Domestic Policy Review of Non-Fuel Minerals", November 1978

OECD "Facing the Future", OECD, Paris, 1979

I.J. *Polmear* "Role of Materials in Energy Production" The Journal of the Australasien Institute of Metals, Vol. 21, Nos. 2 & 3, June–Sept, 1976

Richard W. *Robert*, "Materials Research: A Strategy to Improve the Performance of Materials", in Requirements for Fulfilling a National Materials Policy, ed. by Franklin P. Huddle, Office of Technology Assessment, US Congress, 1974

Barbara *Ward*, "Progress for a Small Planet", W.W. Norton & Company, New York, 1979

Index

Co-author:

T.S. Daugherty
Assistant to the Vice President of Technology, Swiss Aluminium Ltd., Feldeggstrasse 4, 8034 Zurich, Switzerland

Contributors:

Dr. M. B. Bever
Professor of Materials Science and Engineering, Massachusetts Institute of Technology, Cambridge, MA 02139, USA

J. P. Clark
Assistant Professor of Materials Systems, Dept. of Materials Science and Engineering, Massachusetts Institute of Technology, Cambridge, MA 02139, USA

H. Eldag
UNIDO Expert, Main Street 8, 6057 Dietzenbach-Steinberg, West Germany

Dr. G. Friese, Dipl. Chem.
Manager, Staff Group Studies and Perspectives, Swiss Aluminium Ltd., Feldeggstrasse 4, 8034 Zurich, Switzerland

P. Kelterborn, dipl. Ing. ETH SIA
Assistant Vice President, Product Planning, Sika Finanz AG, Rathausstrasse 1, 6340 Baar, Switzerland

I. Reznik, Dipl. Ing.
Manager, Staff Group Studies and Perspectives, Swiss Aluminium Ltd., Feldeggstrasse 4, 8034 Zurich, Switzerland

Dr. D. Spreng
Manager, Staff Group Administration/Ecology, Swiss Aluminium Ltd., Feldeggstrasse 4, 8034 Zurich, Switzerland

F. Tuler
Associate Professor, Materials Science, School of Applied Science and Technology, The Hebrew University of Jerusalem; currently Visiting Scientist, Department of Materials Science and Engineering, Massachusetts Institute of Technology, Cambridge, MA 02139, USA

Acknowledgements

This book contains some ideas and comments of leading experts. I wish to acknowledge and thank the following individuals.

Chapter II

Profiles of 12 Important Metals
Dr. W. Sies, Metallgesellschaft AG, Frankfurt a. Main

Iron and Steel
P. D. Allen, British Steel Corporation, Cardiff
Prof. Dr. L. von Bogdandy, Klöckner & Co., Duisburg
Dr. W. Heinemann, Concast, Zurich
Prof. Dr. F. E. Listhuber, Voest-Alpine AG, Linz
Dr. E. M. Michaelis, Linz
Prof. J. Szekely, MIT, Cambridge, MA
E. Umene, IISI, Brussels

Copper
A. M. Macleod-Smith, Selection Trust Ltd., London
Prof. O.H.C. Messner, Zurich
D. Swan, Kennecot Copper Corp., Lexington, MA

Plastics
Dr. J. K. Craver, Futuresearch, Glendale, MO
Prof. Dr. V. Franzen, Lonza, Basel
Dr. H. Müller-Tamm and Dr. R. Paatz, BASF, Ludwigshafen
Dr. J. Schrade, Alusuisse, Neuhausen

Wood and Wood Products
Jyrki Kettunen and Veikko Vainio, Helsinki
Darrell Ward, Woodworking & Furniture Digest, Wheaton, IL

Advanced Materials
Prof. G. Petzow, Max-Planck-Institut, Stuttgart
Prof. D. Kuhlmann-Wildsdorf and Prof. H. Wilsdorf, University of Virginia, Charlottesville, VA

Chapter VI

Harlan Cleveland, Aspen Institute for Humanistic Studies, New York
Prof. B. Fritsch, ETH, Zurich
Prof. Wm. D. Rowe, American University, Washington, DC
A.P. Sundberg, Consultex S.A., Geneva

The Author

Dieter Altenpohl was born in Barmen, Germany, in 1923. He received a Doctor of Science degree "magna cum laude" in 1948, after studying chemistry at the Technical College in Hannover and physical metallurgy at the University of Göttingen where he was a pupil of Prof. G. Masing.

Since then he has had a broad, varied career spanning three decades, which can be divided into three stages:

1) scientific,
2) process development, and
3) the industry's total environment.

In 1949, he joined the Aluminium Walzwerke Singen GmbH, a subsidiary of Swiss Aluminium Ltd., in West Germany. There he headed a laboratory which carried out scientific industrial research.

He published numerous papers on physical metallurgy, and wrote his first book "Aluminium von innen betrachtet" ("Aluminium Viewed from Within"), Aluminium Verlag Düsseldorf, 1956. This year it will appear in its 4th German edition, 2nd French edition and lst English edition. It has also been translated into Japanese. "Aluminium und Aluminium-Legierungen" ("Aluminium and Aluminium Alloys"), Springer Verlag, Berlin, 1965, was his next book which has become the reference book of the aluminium industry. At the Max-Planck-Institut Stuttgart he was lecturing on the physical metallurgy of light metals.

The beginning of his involvement in process development was the successful development of processes for corrosion protection and for increasing the active surface of electrolytic capacitor electrodes. In 1957, he became head of Operational Research at Consolidated Aluminium Corporation in Jackson, Tennessee, and returned to Europe in 1960 to build up Swiss Aluminium's Process Development Division. In this capacity he was instrumental in pioneering new aluminium-plastic-laminates for structural applications or packaging as well as important new continuous casting processes.

Altenpohl's career entered its third stage in 1970 when he took an increasing interest in the total environment and activities of the materials industries. On this topic he has presented many papers and has lectured at Centre d'Etudes Industrielles in Geneva and the St.Gall Business School. An outgrowth of this work was another book. "TP – Die Zukunftsformel: Möglichkeiten und Grenzen der Technologieplanung" ("TP – The Formula for the Future: Possibilities and Limits of Technology Planning"), Umschau Verlag Frankfurt a. Main, 1975. This book proposes that industry should carry out its own technology assessment for the betterment of the environment.

Currently, Altenpohl is Vice President of Technology at Swiss Aluminium's corporate headquarters in Zurich.

Because he has been active in both industrial and academic circles, he has become an internationally known figure and participates in many international conferences. He has given lectures at the European Management Forum about "Europe's Energy Future" and topics relating to the North-South Dialogue. At preparatory conferences for the 1979 United Nations Conference on Science and Technology for Development (UNCSTD), Altenpohl presented papers on "Assessment of Appropriate Technologies for Emerging Nations" in New Delhi, in 1978 and Mexico City, in 1979.

In recognition of his scientific and academic work to broaden the knowledge of aluminium, Altenpohl was awarded the Sainte-Claire Deville medal by La Société Française de Métallurgie in 1967.

He was Chairman of the Environmental Committee of the International Primary Aluminium Institute from 1974 to 1977. He has been appointed advisor on programs being carried out for the US Congress by the Office of Technology Assessment in the USA. In 1978, the University of Virginia appointed him a Visiting Professor.

MIX
Papier aus verantwortungsvollen Quellen
Paper from responsible sources
FSC® C105338

If you have any concerns about our products,
you can contact us on
ProductSafety@springernature.com

In case Publisher is established outside the EU,
the EU authorized representative is:
Springer Nature Customer Service Center GmbH
Europaplatz 3, 69115 Heidelberg, Germany

Printed by Libri Plureos GmbH
in Hamburg, Germany